AGRICULTURE ISSUES AND POLICIES

SOIL CONTAMINATION

SOURCES, ASSESSMENT AND REMEDIATION

AGRICULTURE ISSUES AND POLICIES

Additional books in this series can be found on Nova's website under the Series tab.

Additional e-books in this series can be found on Nova's website under the eBooks tab.

AGRICULTURE ISSUES AND POLICIES

SOIL CONTAMINATION

SOURCES, ASSESSMENT AND REMEDIATION

JESPER E. LUND
EDITOR

nova science publishers
New York

Library of Congress Cataloging-in-Publication Data

ISBN: 978-1-53613-266-3

Published by Nova Science Publishers, Inc. † New York

CONTENTS

PREFACE

In this compilation, the authors present the results of the systematic study of the spatial distribution of trace metals in surface soil over the Bitola Region, Republic of Macedonia, known for its coal mine and thermoelectrical power plant activities. 149 soil samples were collected including top-soil and bottom-soil samples. Following this, the area in the vicinity of lead–zinc mine "Toranica" near the Kriva Palanka town in Eastern Macedonia was examined for the lithogenic and anthropogenic distribution of 17 elements, and alluvial soil samples and automorphic soil samples were collected from the Kriva Reka River Basin. The authors aim to contribute to their understanding of the behaviour of nanoparticles in the soil, their uptake and distribution within plants, impact of soil microbial communities, and their interactions with other pollutants. Later, the authors summarize a case study on the presence of natural U-isotopes in two soil profiles developed over sandstones from the Paraná sedimentary basin in São Paulo State, Brazil, in order to evaluate the timescale of the weathering processes taking place there. The book suggests synergistic biological mechanisms as a less expensive, environment-friendly method for remediation of pesticide polluted soils. The concluding chapter discusses biochar, a form of char and charcoal made from the partial combustion of organic materials. Its properties include high water-holding capacity, large surface area, cation exchange capacity, and impact, especially on microbial communities.

Chapter 1 - The results of the systematic study of the spatial distribution of trace metals in surface soil over the Bitola Region, Republic of Macedonia, known for its coal mine and thermoelectrical power plant activities are reported. The investigated region (148 km^2) is covered by a sparse sampling grid of 5×5 km, but in the urban zone and around the thermal power plant the sampling grid is denser (1×1 km). In total, 149 soil samples were collected including top-soil (0-5 cm) and bottom-soil samples (20-30 cm). Inductively coupled plasma – atomic emission spectrometry (ICP-AES) was applied for the determinations of 21 elements (Al, As, B, Ba, Ca, Cd, Co, Cr, Cu, Fe, K, Li, Mg, Mn, Na, Ni, P, Pb, Sr, V and Zn). Based on the results of factor analyses, three geogenic associations of elements have been defined: F1 (Fe, Ni, V, Co, Cr, Mn and Li), F2 (Zn, B, Cu, Cd, Na, K) and F3 (Ca, Sr, Mg, Ba and Al). Even typical trace metals such as As, Cd, Cu, Ni, P, Pb and Zn are not isolated into anthropogenic geochemical associations by multivariate statistical methods still show some trends of local anthropogenic enrichment. The distribution maps for each analysed element is showing the higher content of these elements in soil samples collected around the thermoelectric power plants than their average content for the soil samples collected from the whole Bitola Region. It was found that this enrichment is a result of the pollution by fly ash from coal burning which deposited near the plant having a high content of these elements.

Chapter 2 - Anthropogenic activities related to the exploitation of natural resources, technological processing and waste management represent a global problem of environment pollution. Previous studies have shown that certain areas in the territory of the Republic of Macedonia are affected by the anthropogenic introduction of high levels of different chemical elements into the soil. The area in the vicinity of the Toranica lead-zinc mine near the town of Kriva Palanka, eastern Macedonia, was examined regarding the lithogenic and anthropogenic distribution of 17 elements (Al, As, Ba, Ca, Cr, Cu, Fe, K, Li, Mg, Mn, Na, Ni, Pb, Sr, V and Zn). Alluvial soil samples (sediments and fluvisol) and automorphic soil samples (from the topsoil) were collected from the Kriva Reka River basin. The elemental contents were determined by atomic emission spectrometry with inductively coupled plasma (ICP-AES). Data processing was applied with combinations of multivariate statistical methods (factor analysis, principal component analysis and cluster analysis). The summary data for automorphic soil (topsoil and subsoil samples) reveal strong correlations between the elemental distribution in topsoil samples vs. subsoil samples along the Kriva Reka River. Anthropogenic impacts were observed in Zone 1 (the area affected by hydrothermal Pb-Zn exploitation). The soil surface layer was affected by enrichment with As, Mn, Pb and Zn (220, 3400, 1700 and 1100 mg/kg, respectively). Regression analysis revealed significant correlations for Cr, Ni, Pb and Zn in topsoil vs. subsoil (r: 0.92, 0.90, 0.91 and 0.95, respectively).

Chapter 3 - Nanoparticles received great attention due to their unique properties and beneficiary applications. However, bioavailability and toxicity are key features for their massive employment. There are various types of nanoparticles but metal-based nanoparticles (e.g., silver, zinc oxide, titanium dioxide, iron oxide, copper oxide) are widely used in agriculture and allied sectors. It is predicted that soil ecosystems would be the final sink of various types of nanoparticles. So, it is very important to understand the behavior of nanoparticles in the soil and to evaluate the risk of nanoparticles in arable soil ecosystems or other real environmental scenarios. The soil microbiota and plants are among the major co-receptors of nanoparticles; especially, the soil-microbial biomass serves as a pool of nutrients and is a sensitive indicator of microbial changes in soils. Thus, it is more important to explore whether nanoparticles could compromise crop yield, soil properties, soil organisms and functional activities of the soil. Safe and sustainable application of nanotechnology in agriculture requires standards for the applications of nanoparticles in crop plants as well as the soil and water used for agricultural activity. The mechanism of nanoparticles toxicity remains unknown, however, it can be assumed to be closely related to the concentration, structure, size, surface area, and material used for nanoparticles synthesis. Studies conducted in hydroponic conditions do not reflect the interaction of nanoparticles with soil and soil microorganisms. However, real environmental studies should be conducted to observe toxicity of nanoparticles. The chapter will contribute to the better understanding of the fate and behaviour of nanoparticles in the soil, their uptake, and distribution within plants and impact of soil microbial communities as well as their interactions with other pollutants which is an essential prerequisite to environmentally realistic studies of nanoparticles ecotoxicology.

Chapter 4 - Understanding processes involved in the soils formation is important for appropriate assessment of contamination occurring in the superficial horizons. Several tools have been used for such purpose, including the natural radionuclides. Uranium is a lithophile element that is among the main elements contributing to natural terrestrial radioactivity, together with thorium and potassium (^{40}K). The two natural U-isotopes ^{238}U and ^{235}U are precursors of the natural mass number 4n+2 and 4n+3 decay series, respectively. The third natural U-isotope, ^{234}U, is radiogenic and generated in the ^{238}U decay chain according to the sequence: ^{238}U (4.46 Ga, α) $\rightarrow$ ^{234}Th (24.1 days, β^-) $\rightarrow$ ^{234}Pa (1.18 min, α) $\rightarrow$ ^{234}U (248 ka, α) $\rightarrow$ ... ^{206}Pb. If radioactive equilibrium occurs within the ^{238}U decay series, all radionuclides activities are equal. In closed systems, not subjected to weathering processes, ^{234}U reaches radioactive equilibrium with ^{238}U and the $^{234}U/^{238}U$ activity ratio (AR) is unity in the bulk of such systems. ^{234}U needs ~1.2 Ma to reach radioactive equilibrium with ^{238}U in closed systems consisting of many minerals and rocks. Equilibrium and disequilibrium conditions affecting ^{238}U and ^{234}U have been extensively used by the researchers since the 1960s for evaluating the time scale of processes controlling the formation and evolution of soils and weathering profiles. This chapter summarizes a study case focusing the presence of these natural U-isotopes in two soil profiles developed over sandstones from the Paraná sedimentary basin in São Paulo State, Brazil. The reported database has been used in order to evaluate the timescale of the weathering processes taking place there and also for suggesting implications of potential contamination sources into the different soils horizons.

Chapter 5 - Pesticides are generally used for the control of pests and weeds, they are of immense importance and cannot be eradicated, their abuse however may become an issue of concern. Many banned pesticides are still in use under different trade names especially in developing countries due to their active and faster actions and many of such pesticides are grouped as Persistent Organic Pollutants (POPs). POPs usually accumulate in the environment and become persistent with concurrent harmful health hazards. More also, many persistent pesticides are biogenetic in nature posing negative effects on wide range of living organisms at different trophic levels of an ecosystem. Generally, several remediation methods have been suggested for removal of persistent pesticides in the environment 'Bioremediation' (the use of living organisms) is currently gaining attention of biologists and environmental scientists, due to its environment friendly, applicability and cost effective approaches. The use of plants have been well studied for effective remediation of pesticides in soils (Phytoremediation), however, this Bioremediation mechanism is still facing problems such as phytoextraction and phytoaccumulation leading to accumulation of toxic pesticide residues in plant tissues. Plants as primary producers in the ecosystem may transfer toxins into primary consumers then to the secondary consumers, on and on across different food chain. Another problem of Phytoremediation is the difficulty in establishment of plant's growth in heavily polluted soil. In this chapter, synergistic biological mechanisms were suggested as safe, less expensive, environment friendly method for remediation of pesticide polluted soils. The authors are drawing attention to the use of synergistic plant-microbial mechanism as an effective and speedy method of remediation of pesticide-polluted soils feasible at large and small scale. In addition, the authors took insights into the roles played by active enzymatic secretions by soil microorganisms and their genetic inferences during synergistic pesticide degradation. Plant-microbial mechanism relies on symbiotic association between plant root and soil microorganisms leading to exchange of benefits for sustainable adaptation and survival in heavily polluted soils. It also promotes biodegradation and mineralization of pesticides in soil over phytoextraction and accumulation. Many plants and microorganisms can therefore be screened for their synergistic potentials in pesticides removal/degradation in the soil and this is possible with good knowledge of the particular pesticide's nature, its chemical mechanisms, its reactivity combined with adequate knowledge of the plant and microbial genetic and biological nature. These factors are necessary to set-up a sustainable synergistic clean-biotechnology for remediation of a pesticide polluted soil which will be feasible in large and small scale.

Chapter 6 - Biochar is a form of char and charcoal made from the partial combustion of organic materials. Its properties include high water-holding capacity, large surface area, cation exchange capacity, impact especially on microbial communities. These properties are used in agriculture, composting, and land remediation/restoration. Zeolites, natural and synthetic, are microporous aluminosilicate minerals that contain a wide variety of cations, such as Na^+, K^+, Ca^{2+}, Mg^{2+} and others. These ions can easily be exchanged for others in a solution. They have been widely studied regarding their suitability to be used in many different environmental applications worldwide, especially in agriculture. Some of the characteristics of zeolites that make them potentially desirable for improving the properties of soils are a large internal porosity that results in water retention, a uniform particle-size distribution that allows them to be easily incorporated, and a high cation-exchange capacity that enables them to retain nutrients. In literature, there are many published research works relative to the application of biochar to the remediation of soils contaminated with both organic and inorganic contaminants. Heavy metals can be stabilized in soil by adding biochar. It also contains high levels of organic carbon and enriches soils by adding nutrients such as potassium and calcium. Moreover, it increases soil pH, helping neutralize acidic soils, increasing plant and microbial growth, nutrient retention, water holding capacity, preventing groundwater eutrophication and reducing fertilizers requirements. It functions as a filter for the removal of traces of pharmaceutical drug residues and hormones from treated wastewater. Zeolite can be used as a liner for landfills. The ability of zeolites to immobilize contaminants in polluted soils has been also extensively tested. Reducing the plant availability of heavy metals (Cd, Pb, Cr, Zn, Cu, etc.) in soils is critical for optimizing agricultural production in areas with heavy metal contaminated soils. Phytoavailability of heavy metals correlates best with their concentrations in soil solution rather than with their total content in soil. The removal of heavy metals in polluted areas is very difficult because they persist in soils for very long periods. However, the fixation of heavy metals in a non-available form could be a useful method for soils that are already contaminated by heavy metals. Heavy metal phytoavailability may be reduced if the metals are sorbed or precipitated from the soil solution. One of the ways to heavy metal immobilization may be the application of zeolites. Zeolites in general have large cation-exchange capacity and expectedly attract positive-charged ions and, therefore, are widely used for sequestration of cationic pollutants like heavy metals. Natural and artificial zeolites increase ion-exchange sites in soils in addition to offering absorption sites for small molecules, due to their porous structure. Zeolites are able to retain heavy metals in soil. On the other hand, despite promising results for immobilization of contaminants by zeolites, several studies have demonstrated that zeolites addition failed to promote satisfactory growth of plants grown in contaminated soils. Thus, of serious concern within contaminated land remediation should be the possibility of 'Na toxicity' to the surrounding soil, and hence uptake by plants and grazing animals. Consequently, the addition of biochar or zeolite improves crop yields and food quality increasing soil nutrients, adsorbing organic compounds and stabilizing heavy metals in soil.

In: Soil Contamination
Editor: Jesper E. Lund
ISBN: 978-1-53613-266-3

Chapter 1

Geochemical Properties of Topsoil Around the Open Coal Mine and Bitola Thermoelectric Plant, Republic of Macedonia

Trajče Stafilov[1,*], Robert Šajn[2], Mila Arapčeska[3], Ivan Kungulovski[4] and Jasminka Alijagić[2]

[1]Institute of Chemistry, Faculty of Science, Cyril and Methodius University, Skopje, Republic of Macedonia
[2]Geological Survey of Slovenia, Ljubljana, Slovenia
[3]Faculty of Biotechnical Sciences, St. Kliment Ohridski University, Bitola, Republic of Macedonia
[4]Bio Inženering DOO, Skopje, Republic of Macedonia

Abstract

The results of the systematic study of the spatial distribution of trace metals in surface soil over the Bitola Region, Republic of Macedonia,

Corresponding Author Email: trajcest@pmf.ukim.mk.

known for its coal mine and thermoelectrical power plant activities are reported. The investigated region (148 km^2) is covered by a sparse sampling grid of 5×5 km, but in the urban zone and around the thermal power plant the sampling grid is denser (1×1 km). In total, 149 soil samples were collected including top-soil (0-5 cm) and bottom-soil samples (20-30 cm). Inductively coupled plasma – atomic emission spectrometry (ICP-AES) was applied for the determinations of 21 elements (Al, As, B, Ba, Ca, Cd, Co, Cr, Cu, Fe, K, Li, Mg, Mn, Na, Ni, P, Pb, Sr, V and Zn). Based on the results of factor analyses, three geogenic associations of elements have been defined: F1 (Fe, Ni, V, Co, Cr, Mn and Li), F2 (Zn, B, Cu, Cd, Na, K) and F3 (Ca, Sr, Mg, Ba and Al). Even typical trace metals such as As, Cd, Cu, Ni, P, Pb and Zn are not isolated into anthropogenic geochemical associations by multivariate statistical methods still show some trends of local anthropogenic enrichment. The distribution maps for each analysed element is showing the higher content of these elements in soil samples collected around the thermoelectric power plants than their average content for the soil samples collected from the whole Bitola Region. It was found that this enrichment is a result of the pollution by fly ash from coal burning which deposited near the plant having a high content of these elements.

Keywords: soil, trace metals, pollution, thermoelectric power plant, Bitola, Republic of Macedonia

INTRODUCTION

Terrestrial ecosystems and associated fauna and flora are among the principal resources of humans. There is a homeostatic interrelationship between the animate and inanimate environment; however, a significant part of the ecosystem has already been considerably modified by humans, and these processes will continue. Most of the chemical elements necessary for life on the Earth are supplied mainly from the soil overlying the surficial lithosphere. The concentration of trace elements in plants are often positively correlated with the abundance of these elements in the substrate, and the bioavailability of these elements is variable and controlled by specific abiotic and biotic as well as by physical and chemical properties of a given element. Most trace elements essential for humans are also essential for plants. However, the concentration of most elements that may be harmful to humans and animals are not toxic to plants, since they have adapted to such concentrations via increased transfer (Mustafa & Komatsu, 2016, Shahid et al., 2017).

Anthropogenic environmental changes, associated mainly with chemical pollution, lead to a degradation of the natural environment. Among all chemical pollutants, trace elements are of a special ecological, biological and health significance. The production of energy and the consumption of natural resources are the main source of trace elements as contaminants. However, agricultural activities, especially the application of fertilisers and pesticides, also contribute significantly to trace metal pollution in the environment. Many elements are potentially hazardous trace metals to the biosphere (Bowen, 1979). Although some elements such as Cr, Cu, Mn and Zn are essential for plants and humans, they can be hazardous for the environment if released in excessive amounts.

The soil, as a part of the ecosystem, is vital for the survival of mankind which is closely connected to its productivity. Soil functions as a filtering, buffering, storage and transformation system protect against the effects of trace element pollution. The soil is the main source of trace elements for plants, both in terms as micronutrients and as pollutants (Kabata-Pendias & Mukherjee, 2007). The soil is dynamic, open habitats that provide plants with physical support, water, nutrients and air for growth. They also sustain an enormous population of microorganisms, such as bacteria and fungi that recycle chemical elements, notably carbon and nitrogen, as well as toxic substances. The presence of man-made substances in the soil is of great public concern. Many of these chemicals are carcinogens or may accumulate in the environment with toxic effects on ecosystems. Although human exposure to these substances is primarily through food, inhalation or drinking water, soils play an important role because they affect the mobility and biological impact of these toxins.

The abundance of trace metals in the soil has been increased dramatically by the accelerated rate of mineral and fossil fuel extraction and by industrial processes. The problem of ecosystem degradation due to pollution became increasingly acute during the latter decades of the 20th century. Most of the metals were typically found in low concentrations in pristine waters, therefore referred to as trace metals. Rapid increases of trace metal concentrations in the environment are commonly coupled to the development of exploitative technologies.

The soil contains trace elements of various origins: lithogenic elements which are directly inherited from the lithosphere, geogenic elements which are of lithogenic origin, but their concentration and distribution in soil layers and soil particles are changed due to pedogenic processes or anthropogenic elements resulting from direct or indirect anthropogenic activities. The behaviour of trace elements in the soil and, consequently, their bioavailability differs in regards to their origin and depend on soil conditions and chemical and physical properties.

However, regardless of the chemical forms of the anthropogenic trace metals in the soil, their phytoavailability is significantly higher than those of geogenic origin. Soils of several regions of the world (and especially of Europe) have been and will be in the future subject to increased mineral fertilisation, pesticide application, waste disposal and industrial pollution. All these human activities affect both chemical and physical soil properties and lead to changes in the behaviour of trace elements in soils. The impacts of soil acidification, alkalisation, salinity and losses of organic matter on the uptake of trace elements by vegetation, particularly by crop plants, have already become serious environmental and human health issues (Kabata-Pendias & Pendias, 2001; Mirsal, 2008).

Urban soil pollution with trace metals has recently become a subject of many studies, some of which are concerned about large polluted areas (Salminen et al., 2005; Davidson et al., 2006; Mico et al., 2006; Reimann et al., 2011, 2014a, 2014b). Regional contamination of soil occurs mainly in industrial regions and within centres of large settlements where factories, traffic and municipal wastes are the most important sources of trace metals (Kabata-Pendias & Pendias, 2001). Because of heterogeneity and ceaseless changing of urban areas, it is necessary to understand the natural distribution of trace metals and the methods for distinguishing man-made anomalies in nature. The natural background of trace metals content in soil is variable, which means that higher concentrations of some elements can be normal for one region, but anomalous for another.

However, in cases where industrial enterprises, especially mining and metallurgical plants, are situated near cities, pollution can be increased. Recently published papers state that mining and metallurgical activities lead to enormous soil contamination (Kasassi et al., 2008; Aliu et al., 2009, 2010; Stafilov et al., 2010a; Escarré et al., 2011; Šajn et al., 2013; D'Emilio et al., 2013; Li et al., 2014; Modis & Vatalis, 2014;). In this direction very interesting ate the results from the study on the soil pollution due to the coal-fired power plants (Okedeyi et al., 2014; Modis & Vatalis, 2014).

The first results on trace metals environmental pollution in the Republic of Macedonia were obtained from the study on moss biomonitoring of air pollution with trace metals started in 2002. The results of these studies of air pollution in Macedonia indicate decreased air and soil quality in terms of trace metal pollution (Barandovski et al., 2008, 2012, 2013, 2015). The main emission sources are most likely metallurgic activities, power plants and mining activities. Therefore, special studies were performed in the areas with the highest trace metal pollution, including air (moss biomonitoring or attic dust) or soil pollution.

A detailed study was performed to environmental impacts of the Pb-Zn-Cd smelter plant in the city of Veles and detected soil pollution with possibly harmful elements (As, Cd, Cu, Hg, In, Pb, Sb Zn) exceeding the intervention values (Stafilov et. al., 2008a, 2009, 2010b). The study also detected pollution of water and sediment of the Vardar river (Stafilov & Levkov, 2007; Ilić Popov et al., 2014, 2015; Serafimovska et al., 2013). Similar effects were detected in an area with ferronickel metallurgical activities near the city of Kavadarci regarding the contamination of soil (Stafilov et al., 2008b, 2010c) and more significant air pollution with dust rich in Ni, Co and Cr (Bačeva et al., 2011, 2012; Boev et al., 2013; Stafilov et al., 2012, 2014). A detailed study of the impact of the copper mining and flotation plant "Bučim" near the town of Radoviš was also performed, showing the amount of soil (Stafilov et al., 2010d; Balabanova et al., 2013) or air pollution using moss biomonitoring (Balabanova et al., 2009, 2010; Stafilov et al., 2010d, 2014), lichen species biomonitoring (Balabanova et al., 2012, 2014) or attic dust (Balabanova et al., 2009, 2011a, 2011b,

2014a; Stafilov et al., 2012). There are three active Pb-Zn mines and flotation plants situated in the eastern part of the country, and their activities pollute the environment with Pb, Zn, Cd and some other trace metals, which can be found in the soil (Balabanova et al., 2014b, 2015, 2016), the air (Balabanova et al., 2014, 2016, Angelovska et al., 2016), the river water and in sediments (Stafilov & Levkov, 2007; Stafilov et al., 2014a, 2015; Ilić Popov et al., 2014, 2015).

The arsenic-antimony abounded mine "Allchar" near the city of Kavadarci also affects the environment, with the main pollutants As, Sb and Tl in the soil (Bačeva et al., 2014a), the atmosphere (Bačeva et al., 2013), the surrounding vegetation (Bačeva et al., 2014b, 2015) or in river waters and sediments (Stafilov & Levkov, 2007; Stafilov et al., 2013; Bačeva et al., 2013; Ilić Popov et al., 2014). It was concluded that this pollution is a result of both lithological and anthropogenic sources. In the vicinity of the Allchar mine, an open pit mine for nickel ore negatively influences the water and sediments of rivers in the area (Pavlov et al., 2012; Bačeva et al., 2014).

Preliminary studies also found an influence of the thermoelectric power plant near the city of Bitola, producing about one million tons of fly ash per year, deposited on the surface close to the plant. This plant pollutes soils with Pb and Zn (Stafilov et al., 2011, 2014b, 2015; Dimovska et al., 2014a) as well and air with radioactive fly ash containing U and Th (Barandovski et al., 2008) as well as As, Pb and Zn (Dimovska et al., 2014a, 2014b).

Therefore, in the frame of this work, a more detailed study on the trace metal pollution with trace metals in the wide region of Bitola was performed. For that goal in total 229 soil samples were collected and analysed for the assessment of the soil.

Investigated Area

Study area is located in southwest part of the country, in Pelagonia valley (Figures 1 and 2) with an area of 3200 km^2. Thermoelectric power plant "REK Bitola" is a facility of strategic importance and primary installation for electricity generation in the country. It satisfies over 70% of the demand for electricity in the country, as well as major part of the needs for coal (lignite) of industrial boiler plants and general consumption. REK Bitola has been built on the basis of the finds of coal-lignite, at around 12 km eastwards from the city of Bitola. This power plant is major polluter of the environment in the region and it constantly creates large amounts of waste in the form of fly ash and slag (ELEM, 2007).

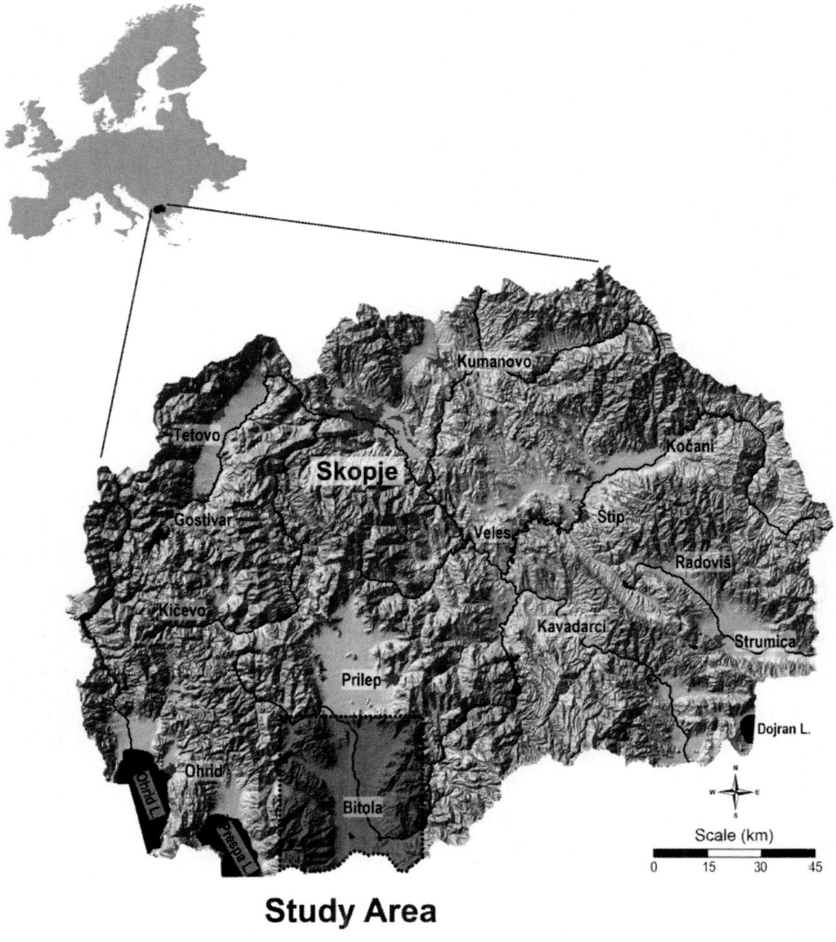

Figure 1. The study area on the territory of the Republic of Macedonia.

There are three municipalities in the study area: Bitola, Novaci and Mogila. The biggest one is Bitola, which is the administrative, cultural, economic, industrial, educational, scientific and diplomatic centre. Bitola is set in the foothills of Mount Baba in middle part of Pelagonia Valley. Through Bitola flows the river Dragor. Bitola is located 14 km north of Greek border at the altitude of 576 meters. Bitola covers the area of 794.53 m^2.

Municipality Novaci covers an area of 773 km^2 and it is located in the southern part of the country at the altitude of 580 meters, at the latitude of 41°2'40" and longitude 21°27'40". Mogila is located in Pelagonia valley, mainly south of the river Šemnica, between the old road Bitola-Prilep (now regional road R-106) and Black River on an average altitude of 583 m. The municipality of Mogila is a middle sized rural municipality, covers an area of 251.2 km^2 and is located in the south-western part of the country.

Pelagonia region has agricultural land with an area of 276,777 ha, from which 117,770 ha are arable agricultural land, and the remaining 159,007 ha are pastures (Figure 2). The structure of arable land - ploughed fields and gardens participate with 83%. As an industry, the agriculture accounts for 4% of production and 5% of total employment in the region. The total forest area in 2014 was 143,545 ha or 14.6% of total forest land. Of ore deposits in the region, most economically important are the coal bearings (lignite) in Pelagonia and Mariovo which today are used in the three blocks of REK Bitola and non-metallic bearings diatomaceous earth in Bitola-Mariovo and white marble in Prilep. The special mineral resource is the cold mineral springs in the village Medjitlija, Bitola. On the territory of Pelagonia region, there are two national parks, Pelister National Park and a part of the National Park Galičica.

Climate and Hydrology

The Bitola Region, as well as the whole Pelagonia Valley, is rather southern positioned and due to the latitude should have an altered Mediterranean climate. However, although Pelagonia Valley is at a distance of 155 km from the Adriatic Sea, and at about 130 km from the Aegean Sea, still the Mediterranean climate influence isn't much felt, because of the high mountainous surrounding of the valley, and its own height above the sea level (it is between 571 and 770 m). That's why the temperate-continental, continental and mountainous climate are mostly felt (Lazarevski, 1993).

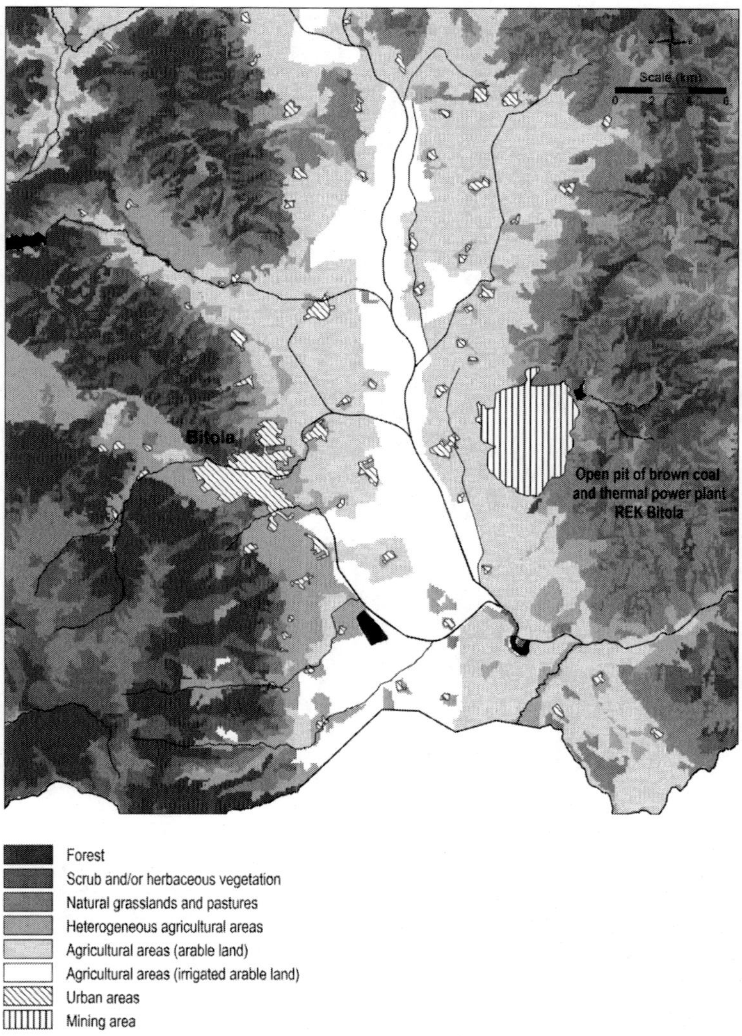

Figure 2. Land-use map of the studying area.

In Bitola the north wind prevails and is with the average annual constancy of 189‰, the average speed is 2.2 m/s and maximum speed is up to 15.5 m/s (Figure 3). The second for its constancy is the south wind with an average of 134‰ with the annual average speed of 3.7 m/s and a maximum speed of 18.9 m/s. The northwest wind is also present with an average of 83‰ with an average speed of 2.4 m/s and maximum speed up to 18.9 m/s. The often drought and relatively high temperatures allow increasing of vapour conditions from free water and soil surface. In this regard, conditions for increasing of possible vapour are created (Lazarevski, 1993).

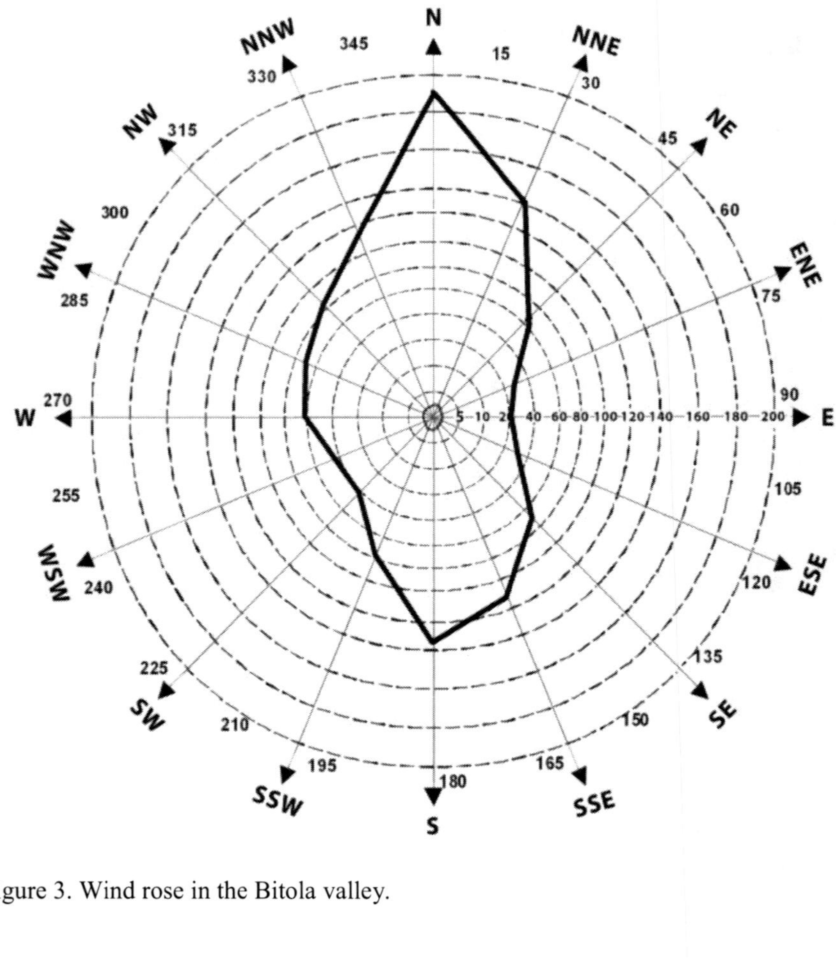

Figure 3. Wind rose in the Bitola valley.

According to the characteristics and the appearance of the climate phenomena during a year, it could be seen that Bitola and Pelagonia belong to a warm continental area. The climate in Bitola has moderate – continental characteristics with an emphasized continental component, because of the closeness of the mountainous relief, the height above the sea level, the nearby valley etc., and these facts make the climate in Bitola and Pelagonia very dynamic and non-stable. The microclimate is influenced by the air pollution caused by the great polluters (REK Bitola, the city boiler rooms, the traffic, the chimneys of the houses and the industrial zone) which release gases that represent one of the components of the thick fogs transiting into industrial smog.

The average rainfall is 640 mm per m^2. The hydrological potential in Pelagonia valley consists of the Crna Reka with its tributaries Blato, Šemnica, Dragor and Eleška River. The catchment of Crna Reka is a part of the Aegean See catchment area.

The basic water resource for watering of agricultural areas in Pelagonia is Streževo Lake. It is also a source of technological water for the needs of REK Bitola. Moreover, there is artificial accumulation Suvodol near REK Bitola.

Geological Description

The study area includes parts of two large tectonic units: the Pelagonian massif and the West-Macedonian zone. The Pelagonian massif is separated from the West-Macedonian zone by a big reverse Pelagonian fault, which is covered by young Quaternary deposits (Stafilov & Šajn, 2016). Simplified geological map with the major lithological unit is provided in Figure 4. The Pelagonian massif is built of Precambrian metamorphic and igneous rocks. The predominant rocks on the western Pelagonian massif are banded muscovite gneisses, but also other varieties such as the banded muscovite-biotite gneisses and Augen-amygdaloidal two-mica gneisses. Medium to coarse-grained granodiorites, garnet-staurolite and garnet-cyanite occur as intrusive bodies in the gneiss-micashist series. By the influence of granodioritic magmas, the surrounding rocks were feldspatized and Augen-amygdaloidal two-mica gneisses and banded two-mica gneisses originated, the feldspatisation processes of muscovite gneiss was weak. Some relicts of amphibolite and schistose amphibolite are present in the study area.

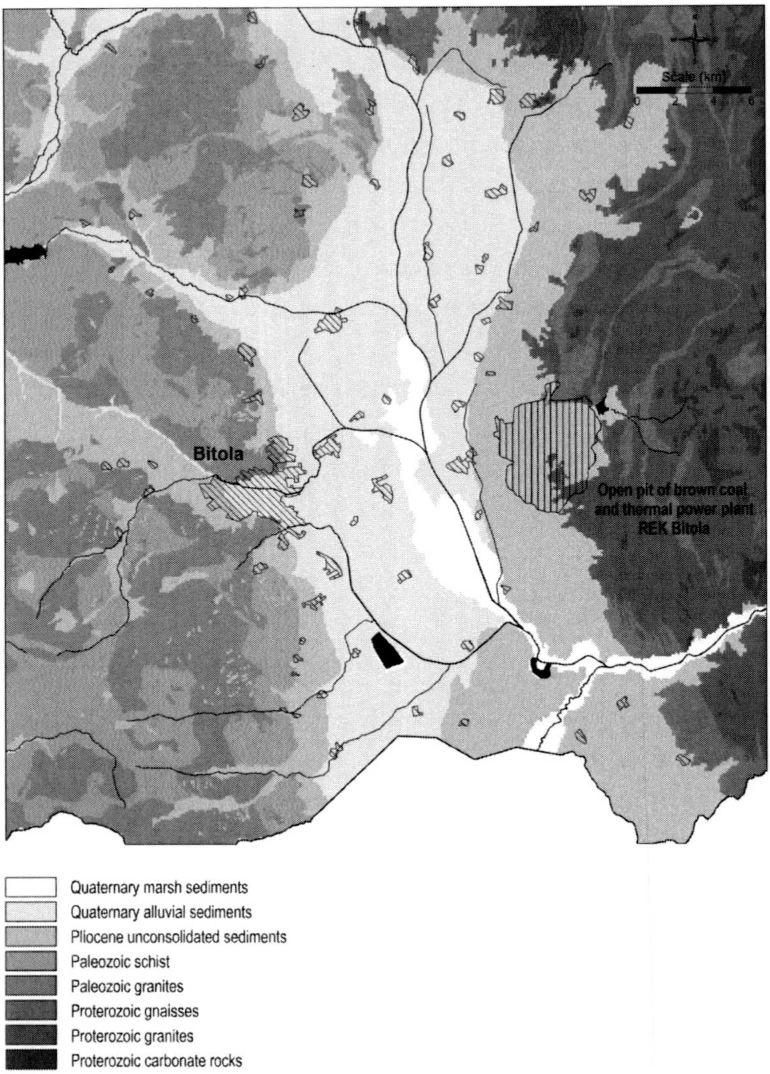

Figure 4. Geological map.

The oldest rocks in the Western Macedonian zone are Paleozoic age and consist of low metamorphic schists and granitic rocks. The most widespread granite is alkalic granite, extending in N-S direction. The oldest granitoid rocks are represented by biotic and amphibolitic granodiorites to which granosyenite and syenite are connected on the southern part of the Western Macedonian zone.

Following the direction N-S, on the left side of the map, Silurian-Devonian formation occur composed of phyllite, slate, metasandstone and metaconglomerate which alter to some older Paleozoic rocks. Ordovician-Silurian greenschist and conglomerates with intrusions of quartz-sericite schist and schistose quartzite developed southern of Bitola.

During Pliocene and Quaternary were first deposited the lacustrine sediments in the Pelagonian depression. They begin with Middle Pliocene gravels, sands and clays with coal beds, while the upper part is composed of poorly sorted gravel, sand and silty clays. Holocene is represented by diluvial and alluvial deposits.

Soil Sampling

For the assessment of soil quality in the Bitola Region, soil samples were collected from 149 locations. From 102 of these locations topsoil (0-5 cm) samples were collected, from 72 locations bottom soil samples from depth 20-30 cm were collected, and from additional 45 locations samples were collected from the depth of 0-30 cm. Also, from 2 locations 10 samples from 5 vertical profiles were collected (0-20 cm, 20-40 cm, 40-60 cm, 60-80 cm 80-100 cm) were collected. Thus, in total 229 soil samples were collected and analysed. The soil sampling locations are presented in Figure 5.

Sampling locations were determined by constructing a network. Depending on site conditions and accessibility, samples were collected in the surroundings of the specified locations, recording the relevant coordinates. To obtain representative composite samples (Darnley et al., 1995), we collected five subsamples from each location on a square plot of

10 × 10 m: four samples at the corners of the plot and one sample at the intersection of the square diagonals.

Sample Preparation and Analysis

The soil samples brought in the laboratory were cleaned from plant material and stones and homogenized, dried at room temperature or in a drying oven at 40°C. Subsequently, they were passed through a 2 mm sieve and ground in a porcelain mortar until reaching a final particle size of 125 μm. Prior to the determination of the elements by atomic emission with inductively coupled plasma (ICP-AES), the samples were chemically prepared by wet digestion, applying a mixture of acids in accordance with the international standards (ISO 14869–1:2001).

All used reagents and standards contained the following analytical purity levels: concentrated nitric acid HNO_3 (69%, ultra-pure, Merck, Germany); concentrated hydrochloric acid, HCl (37%, p.a., Merck, Germany); concentrated hydrofluoric acid, HF, p.a., and concentrated perchloric acid, $HClO_4$, p.a. (Alkaloid Skopje). Redistilled water was used for the preparation of all solutions. The standard solutions for the elements examined were prepared by dissolution of the basic multi-element standard solution with a concentration of 1,000 mg/l (11355–ICP, Multi-Element Standard, Merck). Series of standard solutions were prepared in a linear range in several concentration areas, such as *series 1*: 0,1; 0,2; 0,3; 0,5; 1 mg/l; *series 2*: 1, 2, 3, 5, 10 mg/l and *series 3*: 10, 20, 30, 50, 100 mg/l for the elements in macro-contents.

For each element analysed, previous optimisation of the instrumental conditions was performed. In all samples, the content of a total of 21 elements (Al, As, B, Ba, Ca, Cd, Co, Cr, Cu, Fe, K, Li, Mg, Mn, Na, Ni, P, Pb, Sr, V and Zn) was determined. The quality control was performed by the analysis of certified reference soil and geological samples: soil sample JSAC 0401 (The Japan Society for Analytical Chemistry) and rock CRM samples under saturated igneous rock SARM 3 NIM-L Lujaurite (SA

Bureau of Standards, Pretoria, S. Africa), rock NCS DC71306 (GBW07114) (China National Analysis Centre).

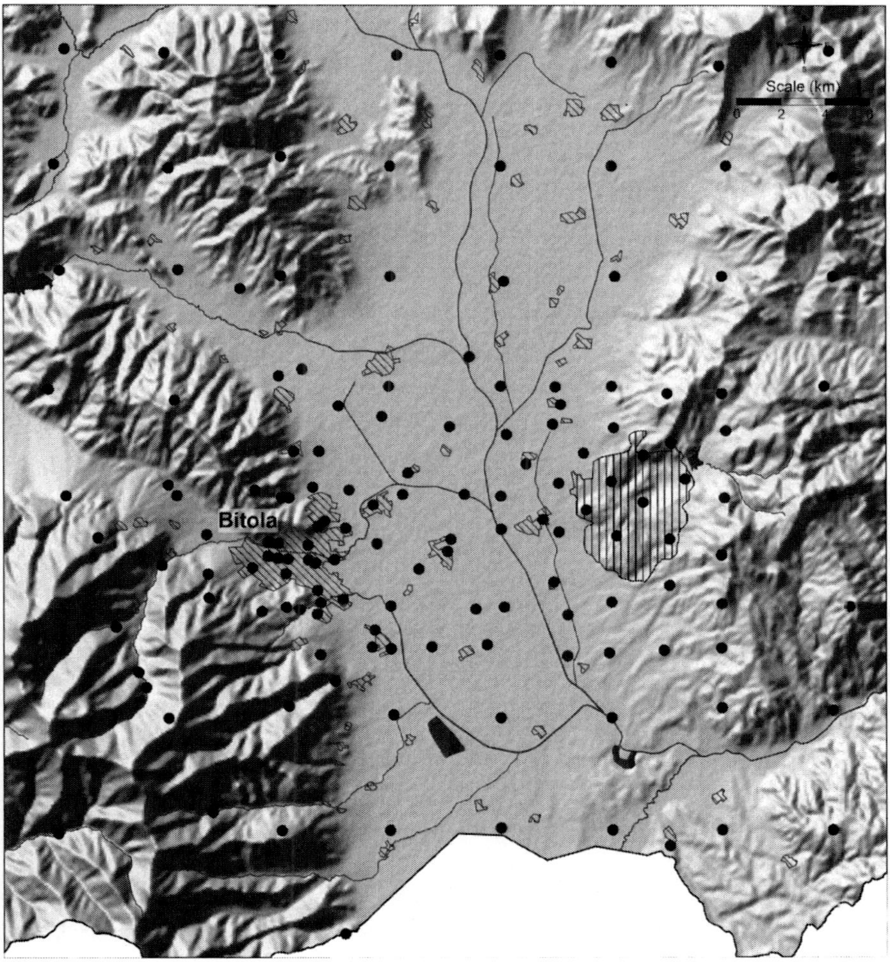

Figure 5. The location in Bitola Region for soil sampling

Data Processing of the Obtained Results

Statistical relationships between two variables can be measured by correlation. Correlations are useful because they can indicate a predictive relationship that can be exploited in practice. A measure of the correlation between two variables is expressed by the correlation coefficient, and the most common is the Pearson correlation coefficient (*r*), which is mainly sensitive to a linear relationship between two variables. Other correlation coefficients have been developed to be more robust than the Pearson correlation, or more sensitive to nonlinear relationships.

Correlation of the values was estimated using bivariate statistics (using the linear correlation of coefficients, (*r*), with significance levels of $p<0.05$ and $p<0.01$. In the case of a large number of variables (for example $n >$ 20), the bivariate analysis is not useful. Therefore, we applied multivariate analysis to enable a reduction of the number of variables by creating new synthetic variables, dependent on the correlation. For processing certain data for comparative analysis, the F-test was applied. We also applied factor analysis (FA) and cluster analysis (CA).

The primary aim of factor analysis is to explain the variation in a set of multidimensional data using as few factors as possible to reveal the hidden data structure (Žibret & Šajn, 2010). In other words, factor analysis is performed on a great number of variables in order to create a small number of new, synthetic variables, called factors. The factors contain large amounts of information about the original variables, and they may have some significance. Factor analysis is performed on variables that are standardised to zero values of a standard deviation unit. Data can be transformed and/or standardised. As a measure of similarity between variables, the product moment correlation coefficient (*r*) was applied. For orthogonal rotation, the varimax method was used.

Cluster analysis (CA) was used with a similar aim as factor analysis, in case high values for the loadings of certain variables were not acquired. In such case is applied the data processing method based on a matrix of remoteness (distances) or nearness (similarities). How the dependence between the variables will be expressed, will depend on the use of a single or several dimensions. To establish the dependence among a series of variables in a multidimensional system, the Euclidean distance matrix is most commonly used.

RESULTS AND DISCUSSION

Data from the descriptive statistics of the measurements of 21 elements determined by ICP-AES (Al, As, B, Ba, Ca, Cd, Co, Cr, Cu, Fe, K, Li, Mg, Mn, Na, Ni, P, Pb, Sr, V and Zn) in all of the topsoil samples (0-5 cm), collected from 102 sampling locations in Bitola Region are presented in Table 1. Data from the descriptive statistics for the content of the same elements in bottom soil samples (20-30 cm and 0-30 cm) collected from 117 sampling locations are given in Table 2. Descriptive statistics of the measurements of these elements in both top and bottom soil samples (229 samples) are given in Table 3. In Tables 1-3 the following statistical parameters are given: N – number of samples; X – arithmetical average; X(BC) – arithmetical average after Box-Cox method; Md – median; Min – minimum; Max –maximum; P_{10} – 10th percentile; P_{90} – 90th percentile; P_{25} – 25th percentile; P_{75} –75th percentile; S – standard deviation; Sx – standard error; CV – coefficient of variation; A – skewness; E – kurtosis; BC – Box-Cox transformed values.

All data were statistically processed using statistic software Stat Soft, 11.0; parametric and nonparametric analyses were applied. Data were subjected to normalisation tests and basic descriptive statistical analysis was conducted on the values of the metal contents in all samples. Distribution histograms were checked visually and the distribution of data on independent variables (elements' contents) was determined.

It could be seen that the median values and ranges of the content of macroelements (Al, Ca, Fe, K, Mg and Na) are in the following ranges: 2.2% and 0.64–6.2% for Al; 0.52% and 0.013–11% for Ca; 2.0% and 0.53–5.2% for Fe; 1.8% and 0.19–3.4% for K; 0.41% and 0.003–1.4% for Mg and 1.1% and 0.067–3.7% for Na. The contents of macroelements are most frequently a result of the dominant geological formations of the area: Quaternary Pliocene unconsolidated sediments, Paleozoic schists and granites and Proterozoic schists, gneisses and granites (Figure 4).

Table 1. Descriptive statistics for the content of the elements in surface soil

Element	Unit	N	X	X(BC)	Md	Min	Max	P10	P90	P25	P75	S	SX	CV	A	E	A(BC)	E(BC)
Al	%	102	2.2	2.1	2.1	0.67	6.2	1.2	3.5	1.6	2.7	0.95	0.075	42	1.06	1.88	0.10	0.03
As	mg/kg	69	21	18	18	2.1	150	6.2	38	13	25	16	1.4	77	4.08	27.84	0.04	1.56
B	mg/kg	84	17	11	18	0.10	97	1.1	33	2.4	27	16	1.3	91	1.58	5.61	-0.38	-1.01
Ba	mg/kg	160	330	280	320	19	1500	71	590	130	460	220	17	67	1.10	3.44	-0.16	-0.52
Ca	%	160	0.80	0.53	0.54	0.019	11	0.15	1.7	0.29	0.97	1.1	0.086	135	6.44	58.18	0.02	1.01
Cd	mg/kg	74	0.46	0.28	0.28	0.10	4.5	0.10	1.0	0.15	0.62	0.52	0.045	112	4.37	29.88	-0.05	-1.18
Co	mg/kg	78	9.4	8.0	8.6	1.2	31	3.3	16	4.9	12	6.0	0.52	64	1.55	3.03	-0.00	-0.17
Cr	mg/kg	102	71	56	56	9.3	650	28	120	42	77	64	5.1	91	5.55	43.69	-0.11	1.82
Cu	mg/kg	102	23	18	20	0.30	250	3.6	41	9.2	32	23	1.8	101	6.18	58.38	0.24	2.57
Fe	%	102	2.2	2.0	2.0	0.53	5.2	1.2	3.5	1.5	2.6	0.90	0.071	42	0.92	0.64	0.02	-0.13
K	%	102	1.8	1.8	1.8	0.19	3.1	1.2	2.5	1.5	2.1	0.49	0.039	27	0.20	0.50	-0.08	0.89
Li	mg/kg	102	17	13	13	1.5	93	4.0	34	6.5	23	13	1.1	80	1.97	6.71	-0.00	-0.62
Mg	%	102	0.43	0.34	0.41	0.0003	1.3	0.030	0.89	0.11	0.64	0.33	0.026	76	0.52	-0.49	-0.28	-0.97
Mn	mg/kg	102	600	540	540	170	2400	310	890	420	680	320	25	53	2.62	10.56	0.12	0.97
Na	%	102	1.2	1.1	1.1	0.067	3.7	0.44	2.0	0.74	1.5	0.62	0.049	53	1.04	1.67	0.01	0.35
Ni	mg/kg	102	26	24	24	1.8	76	9.2	44	16	35	14	1.1	55	0.94	1.18	-0.05	0.10
P	%	102	0.088	0.052	0.050	0.012	1.9	0.025	0.14	0.035	0.083	0.17	0.014	195	8.06	79.2	0.23	0.52
Pb	mg/kg	102	27	10	10	1.0	1500	4.5	34	5.0	20	120	9.7	456	11.87	146.2	-0.28	1.57
Sr	mg/kg	102	89	68	70	3.6	420	16	190	36	120	76	6.0	85	1.72	3.65	-0.01	-0.20
V	mg/kg	102	71	65	65	14	230	34	120	49	90	33	2.6	46	1.25	2.77	-0.13	0.20
Zn	mg/kg	102	96	51	59	1.4	3500	13	150	30	82	280	22	294	11.08	132.2	-0.19	3.08

X = arithmetical average; X(BC) = arithmetical average after Box-Cox method; Md = median; min = minimum; max = maximum; P_{10} = 10th percentile; P_{90} = 90 percentile; P_{25} = 25th percentile; P_{75} = 75th percentile; S = standard deviation; Sx = standard error; CV = coefficient of variation; A = skewness; E = kurtosis; BC = Box-Cox transformed values.

Table 2. Descriptive statistics for the content of the elements in bottom soil

Element	Unit	N	X	X(BC)	Md	Min	Max	P10	P90	P25	P75	S	S_X	CV	A	E	A(BC)	E(BC)
Al	%	117	2.3	2.2	2.2	0.64	4.6	1.2	3.7	1.6	2.9	0.91	0.084	39	0.59	-0.22	-0.05	-0.47
As	mg/kg	96	22	19	18	2.0	76	7.8	37	13	27	13	1.4	63	1.61	3.31	0.05	0.37
B	mg/kg	111	17	8.2	6.8	0.10	180	0.79	36	1.6	29	22	2.1	134	3.95	26.27	0.16	-0.99
Ba	mg/kg	117	350	290	330	14	1500	72	680	130	470	240	23	71	1.19	3.23	-0.13	-0.43
Ca	%	117	0.68	0.44	0.45	0.013	7.6	0.096	1.5	0.23	0.84	0.84	0.078	124	5.13	38.79	-0.11	0.40
Cd	mg/kg	101	0.48	0.24	0.21	0.10	4.3	0.10	1.0	0.10	0.61	0.71	0.071	147	3.67	16.18	0.38	-1.12
Co	mg/kg	106	10	9.2	8.9	1.8	32	4.3	19	6.4	13	5.8	0.56	55	1.23	2.00	-0.04	-0.22
Cr	mg/kg	117	75	61	61	16	540	34	120	45	88	60	5.6	81	4.69	31.96	-0.03	0.90
Cu	mg/kg	117	24	19	20	0.55	250	4.3	45	8.2	31	27	2.5	110	5.73	47.61	0.55	2.55
Fe	%	117	2.3	2.2	2.1	0.82	5.1	1.3	3.6	1.7	3.0	0.93	0.086	40	0.64	-0.06	-0.10	-0.50
K	%	117	1.8	1.8	1.7	0.66	3.4	1.2	2.5	1.4	2.1	0.52	0.048	29	0.37	0.01	0.20	-0.12
Li	mg/kg	117	19	16	17	2.0	94	5.9	34	10	24	14	1.3	72	2.09	7.30	-0.12	-0.03
Mg	%	117	0.43	0.35	0.42	0.0003	1.4	0.029	0.85	0.13	0.65	0.32	0.030	75	0.49	-0.43	-0.34	-0.92
Mn	mg/kg	117	630	560	560	150	2400	300	1000	430	720	330	31	53	2.30	8.28	-0.02	0.82
Na	%	117	1.1	1.0	1.1	0.13	3.3	0.37	2.1	0.68	1.5	0.62	0.058	55	0.73	0.48	-0.12	-0.26
Ni	mg/kg	117	30	27	27	1.0	93	10	52	17	38	17	1.6	58	1.06	1.56	-0.02	0.22
P	%	117	0.082	0.046	0.045	0.008	1.9	0.021	0.11	0.031	0.074	0.18	0.017	224	8.48	81.61	0.01	0.97
Pb	mg/kg	117	25	11	9.6	1.6	410	4.8	39	5.3	21	54	5.0	219	5.31	31.54	0.46	0.28
Sr	mg/kg	117	85	65	64	2.6	370	15	170	37	110	75	6.9	88	1.79	3.58	0.02	-0.04
V	mg/kg	117	79	72	72	20	250	41	130	54	95	39	3.6	49	1.76	5.02	0.09	0.44
Zn	mg/kg	117	110	47	55	0.87	3100	11	140	27	83	390	36	340	7.33	53.72	0.28	3.69

X = arithmetical average; X(BC) = arithmetical average after Box-Cox method; Md = median; min = minimum; max = maximum; P_{10} = 10th percentile; P_{90} = 90 percentile; P_{25} = 25th percentile; P_{75} = 75th percentile; S = standard deviation; Sx = standard error; CV = coefficient of variation; A = skewness; E = kurtosis; BC = Box-Cox transformed values.

Table 3. Descriptive statistics for the content of the elements in surface and bottom soil

Element	Unit	N	X	X(BC)	Md	Min	Max	P10	P90	P25	P75	S	SX	CV	A	E	A(BC)	E(BC)
Al	%	229	2.3	2.1	2.2	0.64	6.2	1.2	3.5	1.5	2.8	0.93	0.059	41	0.84	1.10	-0.01	-0.23
As	mg/kg	176	21	18	18	2.0	150	6.7	37	13	25	16	1.1	73	3.49	22.41	0.04	1.21
B	mg/kg	206	17	9.9	16	0.10	180	0.94	35	2.1	28	19	1.3	112	3.39	23.25	-0.14	-0.99
Ba	mg/kg	245	330	280	320	14	1500	71	600	130	460	230	15	70	1.29	3.89	-0.08	-0.38
Ca	%	245	0.76	0.49	0.52	0.013	11	0.12	1.6	0.24	0.94	1.0	0.066	136	6.09	53.12	0.02	0.70
Cd	mg/kg	186	0.49	0.27	0.26	0.10	4.5	0.10	1.0	0.12	0.64	0.64	0.045	130	3.87	19.54	0.11	-1.22
Co	mg/kg	194	9.9	8.4	8.7	1.2	32	3.6	18	5.5	13	6.2	0.43	63	1.37	2.14	-0.01	-0.30
Cr	mg/kg	229	74	58	57	9.3	650	31	120	42	82	66	4.2	89	4.96	34.90	-0.05	1.27
Cu	mg/kg	229	23	18	20	0.30	250	3.4	42	8.8	32	21	1.4	93	5.50	55.55	0.06	1.47
Fe	%	229	2.2	2.1	2.0	0.53	5.2	1.2	3.6	1.6	2.7	0.94	0.060	42	0.79	0.14	-0.01	-0.36
K	%	229	1.8	1.8	1.8	0.19	3.4	1.2	2.5	1.5	2.1	0.51	0.032	28	0.29	0.27	0.05	0.43
Li	mg/kg	229	17	14	14	1.5	94	4.7	33	7.2	23	14	0.88	79	2.16	7.54	-0.01	-0.41
Mg	%	229	0.43	0.34	0.41	0.0003	1.4	0.027	0.87	0.13	0.65	0.33	0.021	77	0.56	-0.39	-0.27	-0.95
Mn	mg/kg	229	610	550	550	150	2400	310	980	430	700	310	20	51	2.30	8.60	-0.01	0.76
Na	%	229	1.2	1.1	1.1	0.067	3.7	0.42	2.0	0.71	1.5	0.63	0.040	54	0.96	1.23	0.00	0.07
Ni	mg/kg	229	28	25	25	1.0	93	9.2	47	16	36	16	1.0	58	1.07	1.63	0.00	0.14
P	%	229	0.080	0.049	0.048	0.008	1.9	0.023	0.14	0.033	0.076	0.14	0.009	181	9.03	104.9	-0.05	0.72
Pb	mg/kg	229	28	11	10	1.0	1500	4.7	39	5.2	22	110	6.7	382	12.4	173.4	-0.07	1.19
Sr	mg/kg	229	86	66	68	2.6	420	15	180	36	120	73	4.7	85	1.75	3.81	-0.02	-0.14
V	mg/kg	229	74	67	67	14	250	35	120	49	92	37	2.4	50	1.58	4.20	0.00	0.27
Zn	mg/kg	229	110	51	59	0.87	3500	13	160	30	84	350	22	317	8.39	72.41	0.05	3.40

X = arithmetical average; X(BC) = arithmetical average after Box-Cox method; Md = median; min = minimum; max = maximum; P_{10} = 10th percentile; P_{90} = 90 percentile; P_{25} = 25th percentile; P_{75} = 75th percentile; S = standard deviation; Sx = standard error; CV = coefficient of variation; A = skewness; E = kurtosis; BC = Box-Cox transformed values.

Table 4. Comparison of the median, minimal and maximal values for the content of analysed elements in soil from Bitola Region with those for the Republic of Macedonia and Europe

Element	Unit	Bitola Region (this study)		Macedonia (Stafilov & Šajn, 2016)		Europe (Salminen et al., 2005)		Dutch standard*	
		Md	Min–Max	Md	Min–Max	Md	Min–Max	Target	Intervention
Al	%	2.2	0.64–6.2	1.3	0.05–35	5.8	0.70–14.1	–	–
As	mg/kg	18	2.0–150	10	1.0–720	12	0.32–562	29	55
B	mg/kg	16	0.10–180	–	–	–	–	–	–
Ba	mg/kg	320	14–1500	430	6–2900	375	30–1870	200	625
Ca	%	0.52	0.013–11	1.3	0.05–35	0.66	0.019–34.3	–	–
Cd	mg/kg	0.26	0.10–4.5	0.30	0.01–110	0.92	0.03–14	0.8	12
Co	mg/kg	8.7	1.20–32	17	0.5–150	7.78	<3–249	20	240
Cr	mg/kg	57	9.3–650	88	5.0–2700	60	<3–6230	100	380
Cu	mg/kg	20	0.30–250	28	1.6–270	13	0.81–256	36	190
Fe	%	2.0	0.53–5.2	3.5	0.03–12	1.34	0.049–10.6	–	–
K	%	1.8	0.19–3.4	1.9	0.02–5.3	1.59	0.022–5.1	–	–
Li	%	14	1.5–94	26	1.8–210	22	0.14–271	–	–
Mg	%	0.41	0.0003–1.4	0.94	0.12–13	0.47	<0.006–15	–	–
Mn	mg/kg	550	150–2400	900	17–10000	510	31–6070	–	–
Na	%	1.1	0.067–3.7	1.3	0.007–3.7	0.6	0.03–3.34	–	–
Ni	mg/kg	25	1.0–93	46	2.1–2500	18	<2–2690	35	210
P	mg/kg	0.048	0.008–1.9	0.062	0.011–0.39	0.96	0.0082–0.99	–	–
Pb	mg/kg	10	1.0–1500	32	1.2–10000	23	5.3–970	85	530
Sr	mg/kg	6.8	2.6–420	140	21–1400	89	8–3120	–	–
V	mg/kg	67	14–250	89	1.0–470	60	2.7–537	–	–
Zn	mg/kg	59	0.87–3500	83	8.0–10000	52	<3–2900	140	720

Md – median; Min – minimum; Max –maximum.

* https://www.esdat.net/Environmental%20Standards/Dutch/annexS_I2000Dutch%20Environmental%20Standards.pdf.

A comparative analysis was conducted based on the data of the contents of the analysed elements in the soils from Bitola Region and soil from Macedonia (Stafilov & Šajn, 2016) and Europe (Salminen et al., 2005) is given in Table 4. For the comparative analysis, the median values were used as a more stable parameter as well as the range of the content for each element. From data presented in Table 4 it can be seen that the content of Al (2.2%) is higher than the median for the soil from the whole territory of Macedonia, while the contents of the other macroelements (Ca, Fe, K, Li, Mg and Na are lower in relation to the data for Macedonian soil. The median values for all of the trace elements are lower in soil samples from Bitola Region compared with those for Macedonian soil which show that their distribution corresponds to the lithogenic origin of the rocks. Namely, the geology of this region is very specific belonging to the Pelagonian Massif as separate tectonic units in Macedonia (Stafilov & Šajn, 2016). Comparison of the median, minimal and maximal values for the macroelements in soil for the Bitola Region and European soil (Salminen et al., 2005) show lower values for the content of Al, Ca, Mg and Na, and not significant higher value for Fe and K. The content of microelements in soil from Bitola Region is generally similar to those for European soil except for Cd, P, Pb and Sr which are lower in soil from Bitola.

There are no limit values for trace metals in soil in the Republic of Macedonia. Therefore, the obtained data were compared with those of Dutch Standards (http://www.esdat.net/Environmental%20Standards/Dutch/annexS_I2000Dutch%20Environmental%20Standards.pdf). The target and intervention values are given in Table 4. It can be seen that the median values for all elements, besides Ba, are below the target values. Barium content is much higher than the target value of Dutch standards because barium is naturally present in Macedonian soil due to the specific geological formations in this region. However, in some specific parts of the Bitola Region, could be found quite a high content of some trace metals even higher than the intervention value.

Table 5. The ration of the element average content in topsoil (TS) and subsoil (SS)

Elements	Unit	Topsoil	Subsoil	Ratio (TS/SS)
Al	%	2.1	2.2	0.95
As	mg/kg	18	19	0.95
B	mg/kg	11	8.2	**1.34**
Ba	mg/kg	280	290	0.97
Ca	%	0.53	0.44	**1.20**
Cd	mg/kg	0.28	0.24	**1.17**
Co	mg/kg	8.0	9.2	0.87
Cr	mg/kg	56	61	0.92
Cu	mg/kg	18	19	0.95
Fe	%	2.0	2.0	1.00
K	%	1.8	1.8	1.00
Li	mg/kg	13	16	0.81
Mg	%	0.34	0.35	0.97
Mn	mg/kg	540	560	0.96
Na	%	1.1	1.0	1.10
Ni	mg/kg	24	27	0.89
P	%	0.052	0.046	**1.13**
Pb	mg/kg	10	11	0.91
Sr	mg/kg	68	65	1.05
V	mg/kg	65	72	0.90
Zn	mg/kg	51	47	**1.09**

In Table 5 data about the ration of the median element contents in topsoil (TS) and subsoil (SS) samples ate given. As it can be seen the median contents in top and subsoil are very similar and that the ratio of the median contents is lower than 1 for all elements. These findings show in general the absence of the pollution of the topsoil with dust or other source containing a high amount of some specific elements. The exception is the area of REK Bitola were the pollution of soil with some elements present in fly ash deposited in this region influence to the appearance of high content of arsenic, cadmium, lead and zinc. Thus, the content of arsenic in several soil samples collected in the area of REK Bitola and near surroundings is higher than the target value (sample locations: B-7, B-18, MK-63, MK-64, MK-105) or even higher than the intervention value (sample locations: B-18, MK-63, MK-105). The content of cadmium never reaches intervention value but some samples in the area of REK Bitola (sample locations: B5, B12, B14, B16) have slightly increased content over the target value ranging from 1.0 to 1.3 mg/kg. It is interesting that the content of cadmium in the samples from 2 drillings down to 1 m from the fly ash deposit show a high content of cadmium ranging from 1.0 mg/kg in the soil layers from 80-100 cm to 4.5 mg/kg in the top layer of 0-20 cm. Similar results were obtained for the contents of Pb and Zn showing high content in the top layers of fly ash deposit (3070 mg/kg for Zn and 204 mg/kg for Pb).

The matrix of correlation coefficients between all of the analysed elements is given in Table 6. By applying bivariate statistics the degree of correlation between the analysed elements in the soil samples was determined. It was usually assumed that the absolute value of the correlation coefficient between 0.3 and 0.7 indicates a good association and between 0.7 and 1.0 points to the existence of a strong connection between elements. The values for the content of every element were correlated with the values for the content of the other elements. All correlation coefficients between all elements are represented in the matrix of correlation coefficients (Table 6).

In the factor analysis, the distribution is reduced to three synthetic variables (F1 to F3). As it can be seen from the results obtained by factor analysis (Table 7) three factors (i.e., associations of elements) were obtained F1 (Fe, Ni, V, Co, Cr, Mn and Li), F2 (Zn, B, Cu, Cd, Na, K) and F3 (Ca, Sr, Mg, Ba and Al). The total communality of the factors amounts to 72.2%. A total of 21 elements were reduced by factor analysis to 18 elements, which had a total factor loading of 74.4% of the variability (Table 7). The rest of the elements (As, P and Pb) were eliminated from further analysis because they had a low share of communality or low tendency to form independent factors.

The dendrogram obtained by the application of cluster analysis is given in Figure 6. In the dendrogram, the elements are divided into clusters according to the degree of correlation between them. The first cluster includes the elements Al, Mn, Li, Ba, Sr, Mg and Ca, which are included in Factor 3 of the matrix of loading of the dominant rotating factors (Table 7), except for Mn and Li which belong to the association of elements from Factor 1. The second cluster is composed of Co, Cr, Fe, V and Ni. These elements belong to Factor 1 of the factor analysis. The third cluster consists of B, Cd, Cu, Na, Zn and K, which belong to Factor 2).

Table 6. Matrix of correlation coefficients ($n = 160$)

Al	1.00																				
As	0.27	1.00																			
B	-0.23	-0.04	1.00																		
Ba	0.13	0.05	*-0.51*	1.00																	
Ca	*0.50*	0.19	-0.39	*0.53*	1.00																
Cd	-0.21	-0.01	0.70	-0.38	-0.26	1.00															
Co	0.35	0.25	-0.17	0.23	0.20	-0.11	1.00														
Cr	0.28	0.10	-0.26	0.33	0.23	-0.05	*0.68*	1.00													
Cu	0.25	0.19	0.38	-0.14	-0.07	0.35	0.28	0.42	1.00												
Fe	0.48	0.21	-0.12	0.29	0.22	0.02	0.74	0.75	*0.61*	1.00											
K	0.05	-0.07	-0.46	0.40	0.11	-0.43	-0.05	0.02	-0.40	-0.19	1.00										
Li	0.28	0.23	-0.38	0.38	0.01	-0.32	0.39	0.21	0.14	0.40	0.12	1.00									
Mg	0.43	0.12	*-0.51*	0.77	*0.68*	-0.37	0.44	0.45	-0.01	*0.50*	0.27	0.45	1.00								
Mn	0.46	0.25	-0.33	0.30	0.28	-0.22	*0.55*	0.43	0.24	*0.61*	0.12	0.41	0.45	1.00							
Na	0.06	-0.12	-0.46	0.25	0.36	-0.48	-0.16	-0.25	*-0.69*	-0.42	*0.55*	-0.11	0.24	-0.02	1.00						
Ni	0.39	0.16	-0.04	0.24	0.16	0.09	0.71	0.76	*0.61*	0.85	-0.16	0.34	0.42	*0.51*	-0.46	1.00					
P	0.32	0.03	0.20	-0.02	0.13	0.18	0.16	0.23	*0.60*	0.37	-0.07	0.15	0.06	0.25	-0.29	0.35	1.00				
Pb	-0.02	0.10	-0.13	0.24	-0.04	0.16	-0.07	0.22	0.23	0.19	0.12	0.15	0.08	0.10	-0.20	0.17	0.17	1.00			
Sr	0.42	0.07	*-0.56*	0.74	0.82	-0.43	0.25	0.21	-0.26	0.19	0.26	0.20	0.75	0.34	*0.50*	0.16	0.06	-0.03	1.00		
V	0.39	0.12	-0.02	0.24	0.18	0.05	0.72	0.73	*0.61*	0.92	-0.22	0.31	0.41	*0.53*	-0.43	0.84	0.41	0.13	0.16	1.00	
Zn	0.15	0.10	0.37	-0.14	-0.08	*0.51*	-0.02	0.27	*0.65*	0.39	-0.34	-0.01	-0.07	0.08	*-0.56*	0.32	0.36	0.38	-0.28	0.34	1.00
	Al	As	B	Ba	Ca	Cd	Co	Cr	Cu	Fe	K	Li	Mg	Mn	Na	Ni	P	Pb	Sr	V	Zn

Table 7. Matrix of dominant rotated factor loadings (*n* = 260, 18 selected elements)

Element	F1	F2	F3	Comm
Fe	0.90	0.19	0.24	91.1
Ni	0.90	0.22	0.18	88.4
V	0.89	0.22	0.16	87.2
Co	0.84	-0.10	0.07	71.6
Cr	0.81	0.02	0.16	68.3
Mn	0.65	-0.25	0.31	58.1
Li	0.59	-0.38	0.16	51.7
Zn	0.07	0.82	0.09	69.3
B	-0.23	0.79	-0.27	75.0
Cu	0.46	0.71	-0.01	71.3
Cd	-0.08	0.65	-0.27	50.2
Na	-0.31	-0.68	0.39	70.8
K	0.02	-0.69	0.15	50.2
Ca	0.08	-0.05	0.93	87.4
Sr	0.18	-0.36	0.86	89.8
Mg	0.45	-0.28	0.75	84.3
Ba	0.27	-0.48	0.67	75.5
Al	0.44	0.02	0.63	58.8
Prp. Totl	30.2	22.0	20.0	72.2
Eigen Val	6.93	4.58	1.48	
Expl. Var	5.43	3.96	3.60	

F1, F2, F4 – Factor loadings; Var – Variance (%); Com – Communality (%).
Prp. Tol – total amount of the explained system variance.
Expl. Var – particular component variance; Eigen Value – Eigene value.

To construct the maps for the distribution of the elements as well as for the factor associations, the universal kriging method was applied. Kriging is an optimal prediction method, indented for geophysical variables with continuous data distribution. The obtained values of variables can sometimes be random, but their variance is not described with geometric function. This method performs projection of an object by using the values of certain parameters that describe its position (latitude, longitude and oval height), i.e., any object can be spatially defined (Kaymaz, 2005). All data about variables are organized in several possible ways; most common one used is raster. Generally, raster consists of a matrix of cells (pixels) arranged in rows and columns, where each cell contains values that provide information about variables. Digital air–photography, satellite imagery, digital images and scanned maps can be used as a raster. Kriging interpolation performs the output of each raster cell by calculating the average load of nearby vectors. Kriging method analyses the statistical variation of the values of different distances and at various positions, determines the shape and size of the specified point for examination as a set of load factors (Sakata et al., 2003).

For the development of the distribution maps, the kriging method with linear variogram interpolation was applied. As area limits, we considered the percentile values of the distribution of the interpolated values. The following seven areas of percentile values were selected: 0–10, 10–25, 25–40, 40–60, 60–75, 75–90 and 90– 100. For a graphical display of spatial distribution, the maps with percentile distribution have been used, where different colours represent different concentration arrangements. The distribution maps for the individual elements are prepared in 7 concentration ranges for each element.

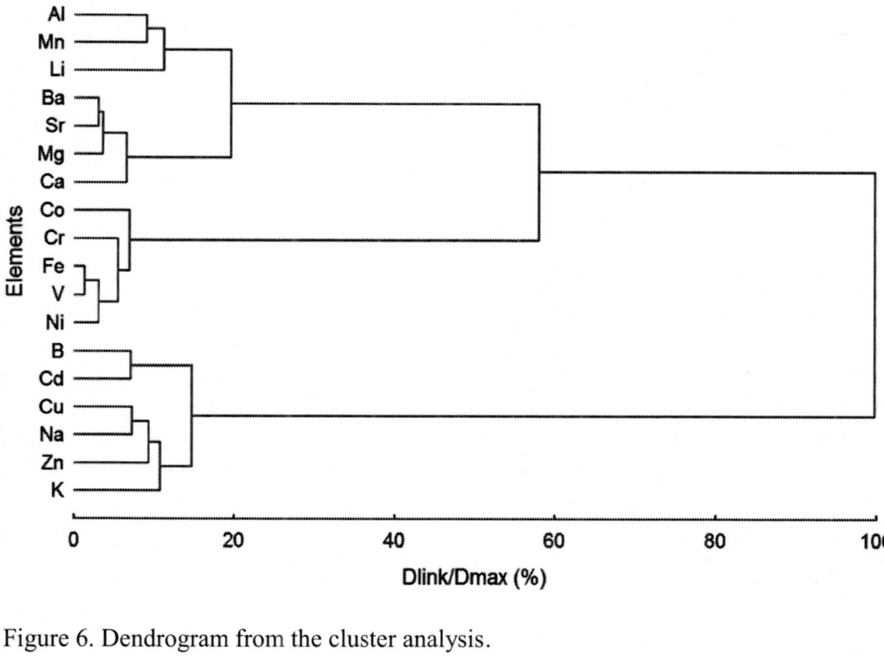

Figure 6. Dendrogram from the cluster analysis.

Table 8. The average content of the analysed elements in soil and fly ash

Element	Unit	Soil	Fly ash
Al	%	2.1	2.2
As	mg/kg	18	**31**
B	mg/kg	9.9	36
Ba	mg/kg	280	127
Ca	mg/kg	0.49	0.42
Cd	mg/kg	0.27	**2.3**
Co	mg/kg	8.4	5.8
Cr	mg/kg	58	47
Cu	mg/kg	18	**42**
Fe	%	2.1	2.1
K	%	1.8	0.87
Li	mg/kg	14	31
Mg	%	0.34	0.20
Mn	mg/kg	550	322
Na	%	1.1	0.14
Ni	mg/kg	25	**32**
P	%	0.049	**0.11**
Pb	mg/kg	11	**102**
Sr	mg/kg	66	22
V	mg/kg	67	61
Zn	mg/kg	51	**1002**

Factor 1 (Fe, Ni, V, Co, Cr, Mn and Li)

The spatial distribution map of the factor scores of factor F1 (Figure 7) shows that these elements are present in varying concentrations throughout the studied area. Higher contents of these elements are found in the soil samples from the southern part (Mts. Baba and Nidže), in some part of the Selečka Mountain to the eastern part and in the central part of the studied the area. According to geological formations, higher content of these elements are present in the soil from the areas were Pliocene sediments and Paleozoic and Proterozoic schists are dominant (Figure 7). The distribution maps for all of the elements included in F1 show that their distribution follows the distribution of factor values of F1. In general, it could be concluded that the distribution of the elements from Factor 1 is, their highest contents are found in soil over Mesozoic and Paleozoic magmatic rocks from Pelagonian massif, but they are characterised by their almost complete absence in the Pelagonian massif (Precambrian metamorphic and carbonate rocks) (Stafilov & Šajn, 2016). From the data presented in Table 8, it could be seen that the content of the elements in soil samples from the Bitola area included in Factor 1 are a bit higher or very close to their content in fly ash samples confirming their lithological origin.

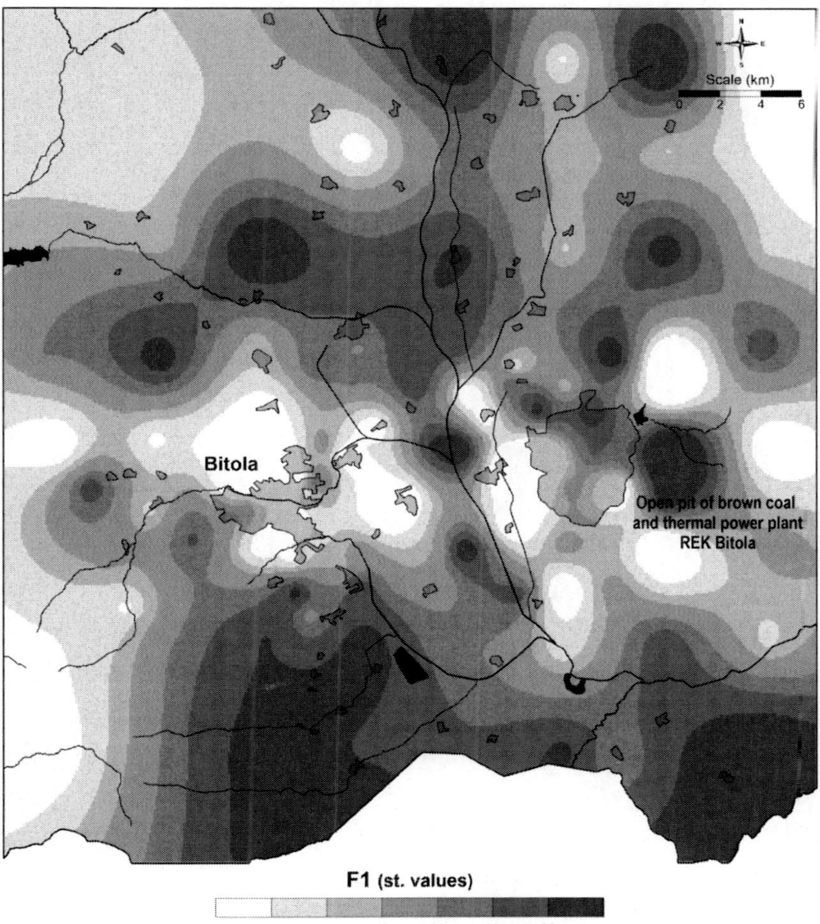

Figure 7. Spatial distribution of factor scores of F1 (Fe, Ni, V, Co, Cr, Mn and Li).

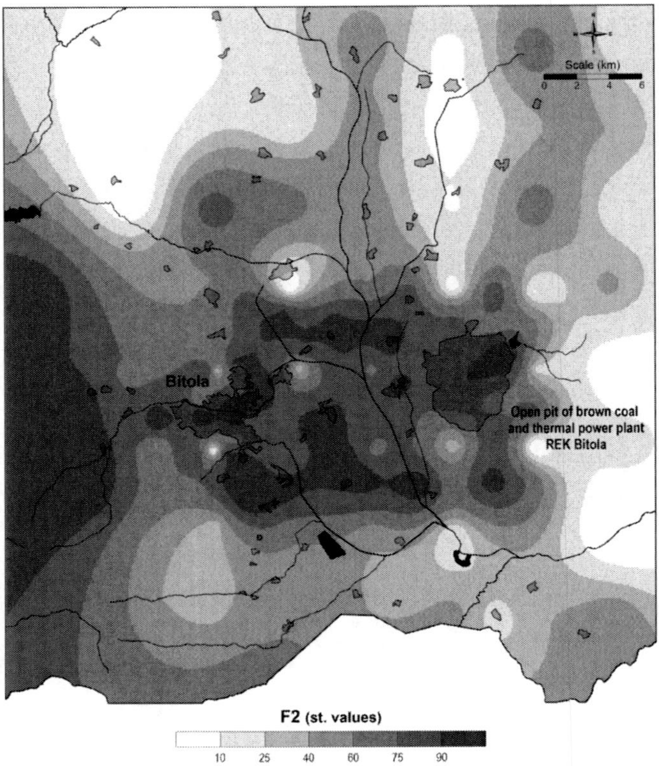

Figure 8. Spatial distribution of factor scores of F2 (Zn, B, Cu, Cd, Na, and K).

F2 (Zn, B, Cu, Cd, Na and K)

The spatial distribution map of the factor scores of factor F2 (Figure 8) shows that these elements are also present in varying concentrations throughout the studied area. Higher content of the elements associated with this factor was found in the soil samples from the mountain Baba as well as in the area of the valley between the city of Bitola and REK Bitola. According to geological formations, higher content of these elements are present in the soil from the areas were Paleozoic and Proterozoic schists and granites and Quaternary sediments. It could be seen that the distribution of each element from this association follows the distribution factor values F1. From data presented in Table 8 and from the distribution maps for the elements included in this association it could be concluded that Factor 2 represents a geogenic-anthropogenic association Namely the distribution of Na and K follows the geological formation and their content is much lower in fly ash samples that in the soil from all region (Figures 9). However, the average content of B, Cd, Cu and Zn is several times higher in fly ash than in their average content in soil from the region (Table 8). Thus the content of B, Cd, Cu and Zn in fly ash is 36, 2.3, 42 and 1002 mg/kg, and 9.9, 0.27, 18 and 51 mg/kg in fly ash, respectively. This influence could be seen also from the distribution maps of these elements (Figures 10-11) with increased content of these elements in the soil in the area around the thermoelectric power plant and the city of Bitola.

F3 (Ca, Sr, Mg, Ba and Al)

The spatial distribution map of the factor scores of factor F3 (Figure 12) shows that these elements are also present in varying concentrations throughout the studied area. Higher content of the elements associated with this factor was found in the soil samples from the mountain Baba and southern part of the Pelagonia Valley. According to geological formations, higher content of these elements are present in the soil from the areas were Quaternary and Pliocene unconsolidated sediments, as well as Paleozoic schists and granites, are dominant geological formations (Figure 4). The maps of spatial distribution of all elements included in Factor 1 are very similar following the geological formations. Arsenic, phosphorous and lead are not included in the factors due to the absence of correlation with the other elements. The maps of spatial distribution of these elements are presented in Figures 13-14. From Figure 13 it could be seen that the content of arsenic in soils from the investigated area is relatively low (average value of 18 mg/kg) while the average content of arsenic in fly ash is 31 mg/kg. However, the content of arsenic in several soil samples collected in the area of REK Bitola and near surroundings are higher than target value according to the Dutch value (29 mg/kg) or even higher than the intervention value of 55 mg/kg. This is the results of the pollution with fly ash in the near surrounding of the REK Bitola plant (Table 8).

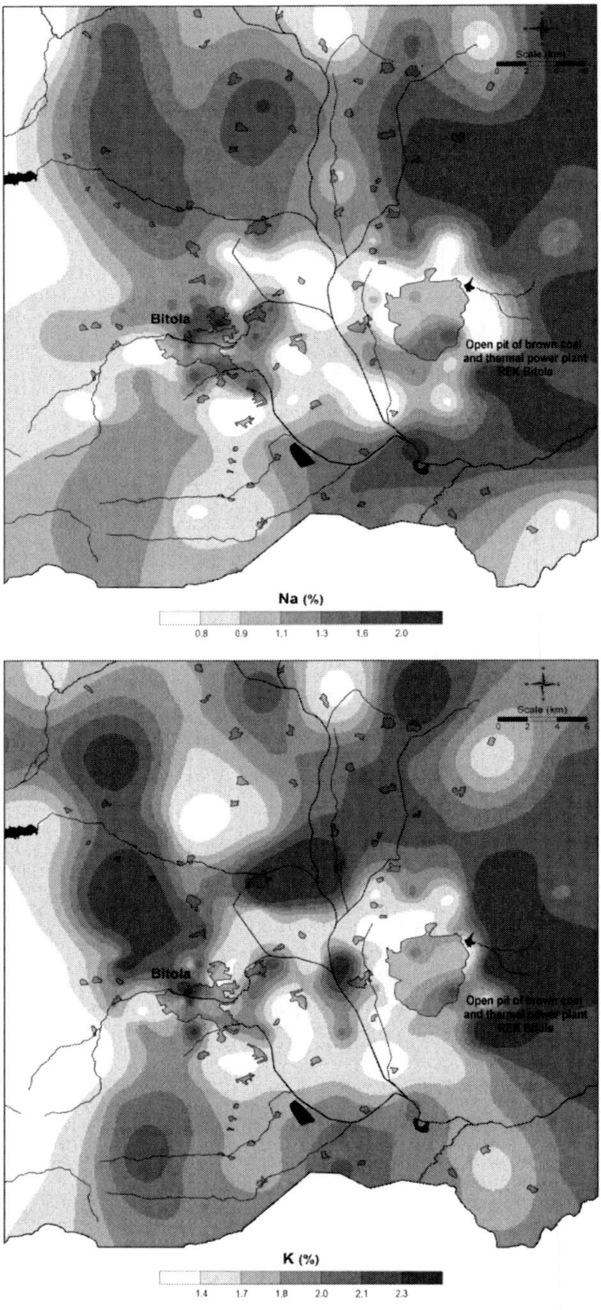

Figure 9. Spatial distribution of sodium (above) and potassium (below).

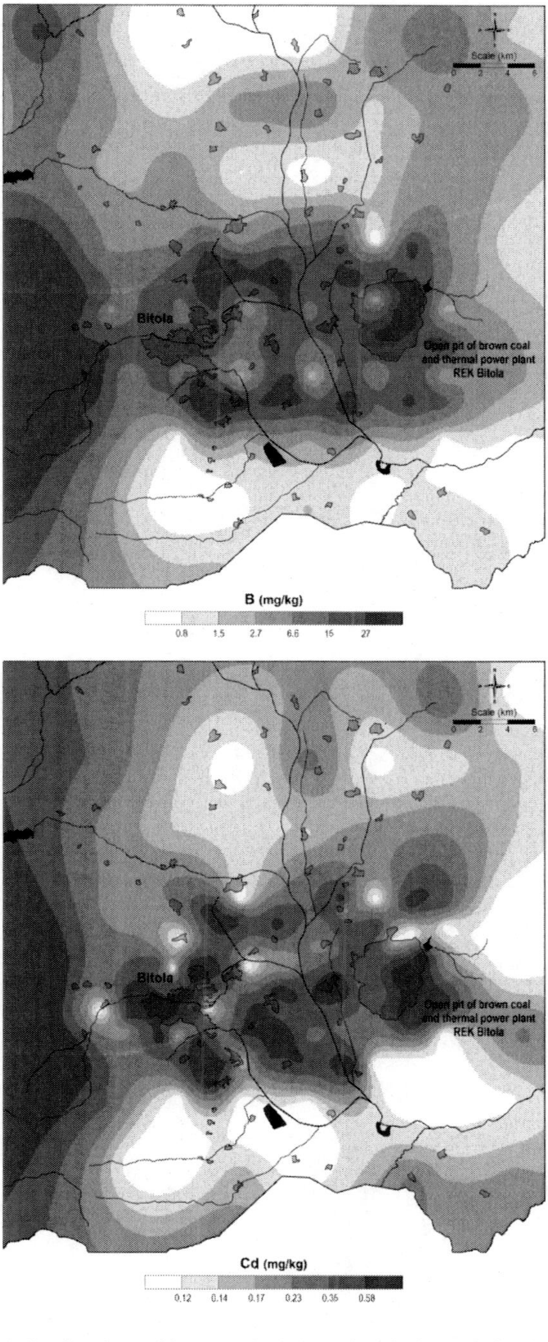

Figure 10. Spatial distribution of boron (above) and cadmium (below).

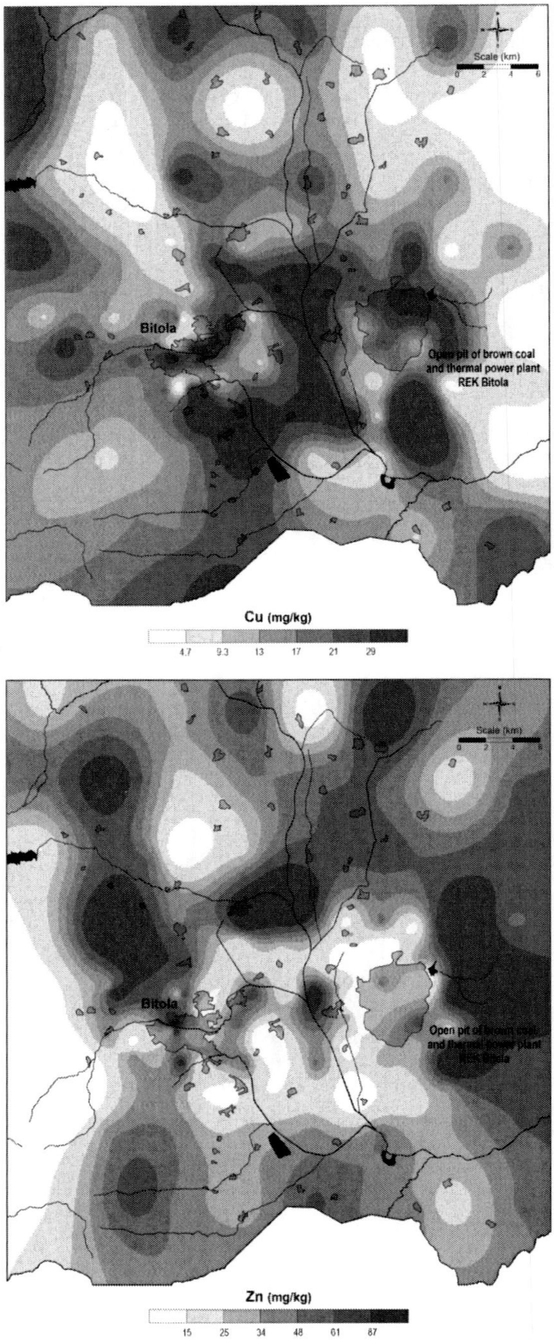

Figure 11. Spatial distribution of copper (above) and zinc (below).

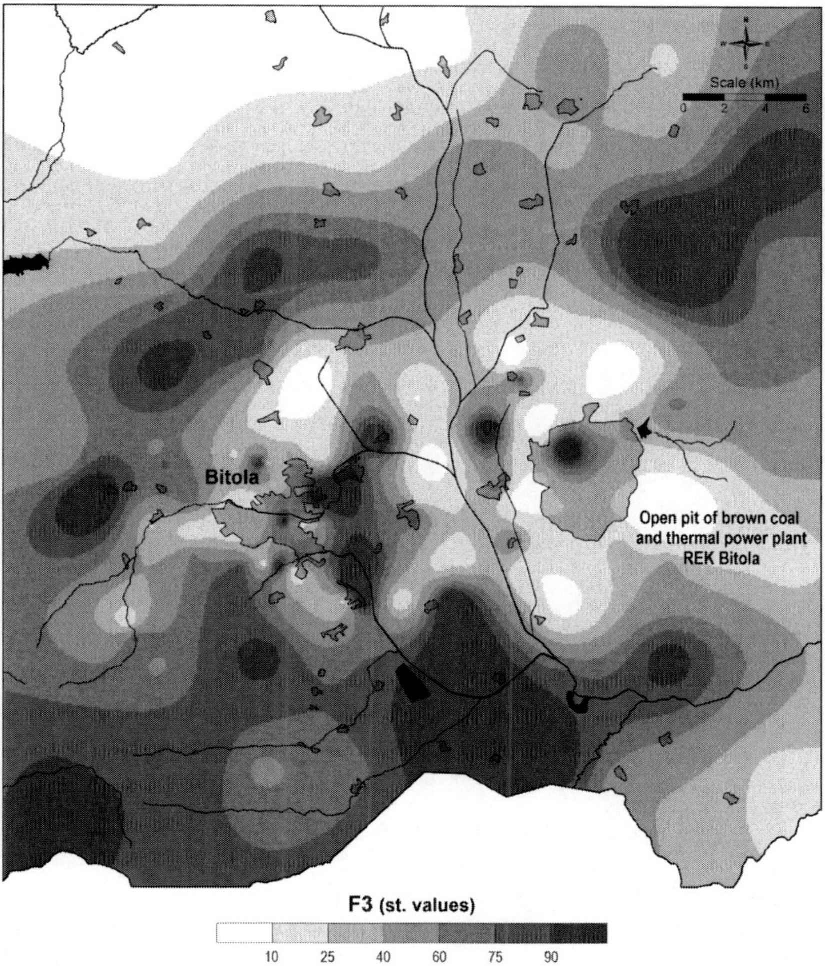

Figure 12. Spatial distribution of factor scores of F3 (Ca, Sr, Mg, Ba and Al).

Spatial distribution of phosphorous presented in Figure 14 shows that higher content of P (average value of 0.049%) are found in soil samples collected from the agricultural land of Pelagonia Valley. It was also found that the content of P is higher in soil collected in the vicinity of REK Bitola (average of 0.11%). In some samples, the content of phosphorous is higher than 1%. Such high values are the most probably due to phosphate fertilizers used in the agricultural practice in this region.

The content of lead in the whole investigated area is relatively low ranging from 1 to 1500 mg/kg with an average value of 28 mg/kg and the median value of 10 mg/kg (Table 3). The higher content was found in the soil samples collected in the city of Bitola and in its near vicinity as well as in the area of REK Bitola (Figure 14). The reason for higher content in these areas is probably the urban activities in the city of Bitola and the pollution in the area of REK Bitola due to higher content of lead in fly ash (from 20 to 258 mg/kg with an average of 102 mg/kg) (Table 8).

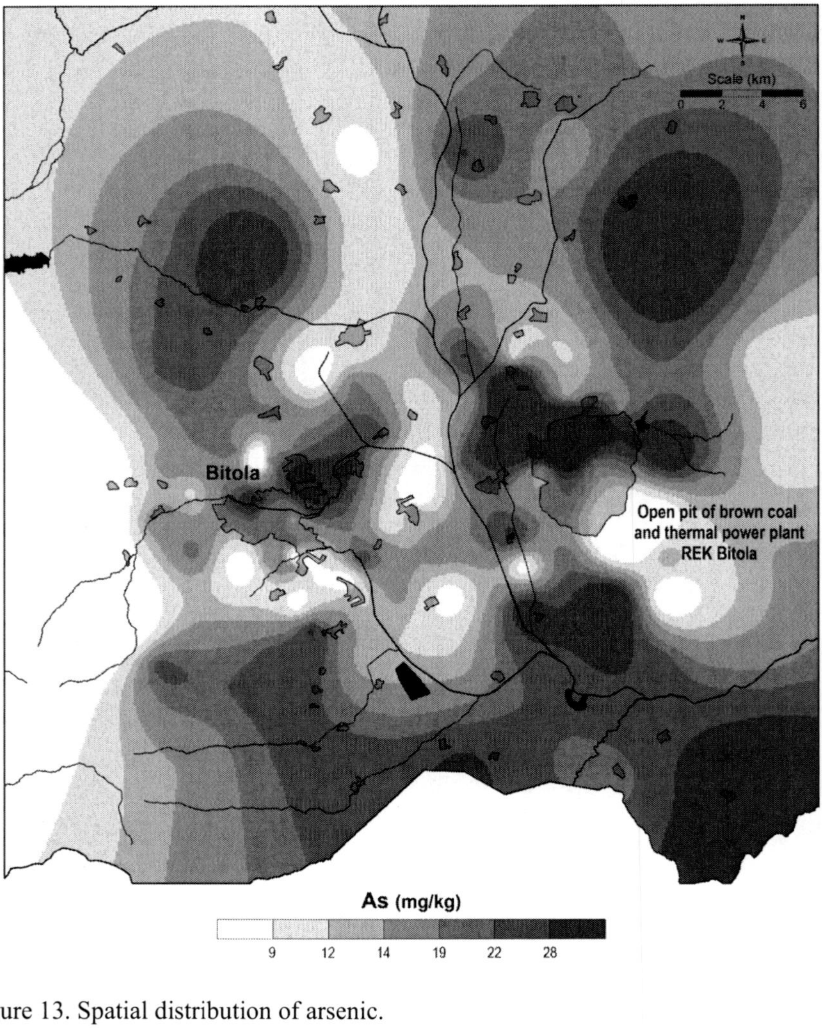

Figure 13. Spatial distribution of arsenic.

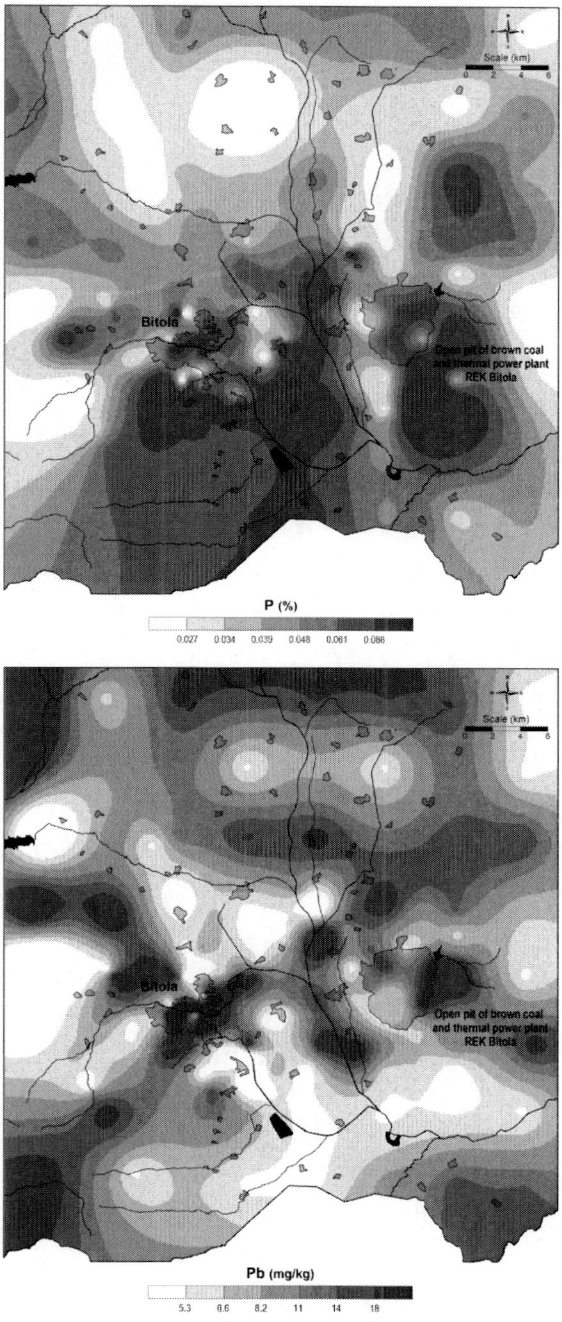

Figure 14. Spatial distribution of phosphorous (above) and lead (below).

CONCLUSION

From the obtained results it could be concluded that the content of soil in the investigated area corresponds to the lithological characteristics of the Bitola Region in the Southern Pelagonia. The factor analysis gives four associations of elements: F1 (Fe, Ni, V, Co, Cr, Mn and Li), F2 (Zn, B, Cu, Cd, Na, and K) and F3 (Ca, Sr, Mg, Ba and Al). The higher content of the elements included in Factor 1 (Fe, Ni, V, Co, Cr, Mn and Li) are found in the soil samples from the southern part (Mts. Baba and Nidže), in some part of the Selečka Mountain to the eastern part and in the central part of the studied the area. The distribution maps for all of the elements included in F1 show that their distribution follows the distribution of factor values of F1. In general it could be concluded that the distribution of the elements from Factor 1 is, their highest contents are found in soil over Mesozoic and Paleozoic magmatic rocks from Pelagonian massif, but they are characterized by their almost complete absence in the Pelagonian massif (Precambria metamorphic and carbonate rocks) (Stafilov & Šajn, 2016).

The spatial distribution map of the factor scores of factor F2 (Zn, B, Cu, Cd, Na, and K) shows that these elements are also present in varying concentrations throughout the studied area. Higher content of the elements associated with this factor was found in the soil samples from the mountain Baba as well as in the area of the valley between the city of Bitola and REK Bitola. According to geological formations, higher content of these elements are present in the soil from the areas were Paleozoic and Proterozoic schists and granites and Quaternary sediments. It could be also seen that the distribution of each element from this association follows the distribution factor values F1. The only exception is the distribution of copper and nickel because their higher contents are found in the area of fly ash deposit of REK Bitola (from 40-65 mg/kg, 25 to 93 mg/kg, respectively) due to the pollution by fly ash reach in copper.

The spatial distribution map of the factor scores of factor F3 (Ca, Sr, Mg, Ba and Al) shows that these elements are also present in varying concentrations throughout the studied area. Higher content of the elements associated with this factor was found in the soil samples from the mountain

Baba and southern part of the Pelagonia Valley. According to geological formations, higher content of these elements are present in the soil from the areas were Quaternary and Pliocene unconsolidated sediments, as well as Paleozoic schists and granites, are dominant geological formations. It could be seen that the distribution of each element from this association follows the distribution of factor values of F3.

The map of the spatial distribution of arsenic shows that its content in soils from the investigated area is relatively low. However, the content of arsenic in several soil samples collected in the area of REK Bitola and near surroundings is higher than target value according to the Dutch value (29 mg/kg) or even higher than the intervention value of 55 mg/kg. The content of lead in the whole investigated area is relatively low with an average value of 28 mg/kg and the median value of 10 mg/kg. The higher content was found in the soil samples collected in the city of Bitola and in the near vicinity of REK Bitola. The reason for higher content in these areas is probably the urban activities in the city of Bitola and the pollution in the area of REK Bitola due to higher content of lead in fly ash (from 20 to 258 mg/kg with an average of 110 mg/kg).

Spatial distribution of phosphorous shows that higher content of P in soil samples collected from the agricultural land of Pelagonia Valley. It was also found that the content of P is higher in soil collected in the vicinity of REK Bitola. In some samples, the content of phosphorous is higher than 1%. Such high values are the most probably due to phosphate fertilizers usage in the agricultural practice in this region.

ACKNOWLEDGMENTS

Research in this Chapter was realized as part of the project "Soil Degradation Assessment and Rehabilitation Strategies for Sustainable Land use planning" - TERRA MED - IPA/2010/DN023658/CN362856, founded by the Europe Union and by Ministry of Local Self-Government of the Republic of Macedonia under the IPA Cross Border Programme with Greece and part of the research was financially supported by the

Slovenian Research Agency (research core funding No. P1-0020 - Groundwaters and geochemistry).

References

Aliu, M., Šajn, R., Stafilov, T., (2009). Distribution of cadmium in surface soils in K. Mitrovica Region, Kosovo. *Geologica Macedonica*, 23, 27–34.

Aliu, M., Šajn, R., Stafilov, T., (2010). Distribution of zinc in surface soils in K. Mitrovica Region, Kosovo. *International Journal of Pure and Applied Chemistry*, 5(4), 351-357.

Angelovska, S.,Stafilov, T.,Šajn, R.,Balabanova, B., (2016). Geogenic and anthropogenic moss responsiveness on lithological elements distribution around Pb-Zn ore deposit. *Archives of Environmental Contamination and Toxicology*, 70, 487-505.

Bačeva, K., Stafilov, T., Šajn, R., Tănăselia, C., Ilić Popov, S., (2011). Distribution of chemical elements in attic dust in the vicinity of ferronickel smelter plant. *Fresenius Environmental Bulletin*, 20(9), 2306–2314.

Bačeva, K., Stafilov, T., Šajn, R., Tănăselia, C., (2012). Moss biomonitoring of air pollution with heavy metals in the vicinity of a ferronickel smelter plant. *Journal of Environmental Science and Health, Part A*, 47(4), 645–656.

Bačeva, K., Stafilov, T., Šajn, R., Tănăselia, C., (2013). Air dispersion of heavy metals in vicinity of the As-Sb-Tl abounded mine and responsiveness of moss as a biomonitoring media in small scale investigations. *Environmental Science and Pollution Research*, 20(12), 8763–8779.

Bačeva, K., Stafilov, T., Šajn, R., Tănăselia, C., Makreski, P., (2014a). Distribution of chemical elements in soils and stream sediments in the area of abandoned Sb-As-Tl Allchar mine. Republic of Macedonia. *Environmental Research*, 133, 77–89.

Bačeva, K., Stafilov, T., Matevski, V., (2014b). Bioaccumulation of heavy metals by endemic *Viola* species from the soil in the vicinity of the As-Sb-Tl mine "Allchar", Republic of Macedonia. *International Journal of Phytoremediation,* 16, 347-365.

Bačeva, K., Stafilov, T., Matevski, V., (2015). Distribution and mobility of toxic metals in *Thymus alsarensis* Ronniger in the Allchar As-Sb-Tl mine, Republic of Macedonia. *Plant Biosystems*, 149(5), 884–893.

Balabanova, B., Stafilov, T., Bačeva, K., Šajn, R., (2009). Atmospheric pollution with copper around the copper mine and flotation, Bučim, Republic of Macedonia, using biomonitoring moss and lichen technique. *Geologica Macedonica*, 23, 35–41.

Balabanova, B., Stafilov, T., Bačeva, K., Šajn, R., (2010). Biomonitoring of atmospheric pollution with heavy metals in the copper mine vicinity located near Radoviš, Republic of Macedonia. *Journal of Environmental Science and Health, Part A*, 45, 1504–1518.

Balabanova, B., Stafilov, T., Šajn, R., Bačeva, K., (2011a). Distribution of chemical elements in attic dust as reflection of lithology and anthropogenic influence in the vicinity of copper mine and flotation. *Archives of Environmental Contamination and Toxicology*, 61(2), 173–184.

Balabanova, B., Stafilov, T., Šajn, R., Bačeva, K., (2011b). Total deposited dust as a reflection of heavy metals distribution in area with intensively exploited copper minerals. *Geologica Macedonica*, 25(1–2), 1–9.

Balabanova, B., Stafilov, T., Šajn, R., Bačeva, K., (2012). Characterisation of heavy metals in lichen species *Hypogymnia Physodes* and *Evernia Prunastri* due to biomonitoring of air pollution in the vicinity of copper mine. *International Journal of Environmental Research*, 6(3), 779-792.

Balabanova, B., Stafilov, T., Šajn, R., Bačeva, K., (2013). Spatial distribution and characterization of some toxic metals and lithogenic elements in topsoil and subsoil from copper mine environs. *International Journal of Environmental Protection*, 3(9), 1–9.

Balabanova, B., Stafilov, T., Šajn, R., Bačeva, K., (2014a). Comparison of response of moss, lichens and attic dust to geology and atmospheric pollution from copper mine. *International Journal of Environmental Science and Technology*, 11, 517-528.

Balabanova, B., Stafilov, T., Šajn, R., (2014b). Variability assessment for multivariate analysis of elemental distributions due to anthropogenic and geogenic impact in lead-zinc mine and flotation environ (moss sampling media). *Geologica Macedonica*, 28(2), 35-41.

Balabanova, B., Stafilov, T., Šajn, R., Tănăselia, C., (2015a). Lead distribution in soil due to characteristic lithogenic and anthropogenic factors in the Bregalnica River Basin. *Geologica Macedonica*, 29, 53-61.

Balabanova, B., Stafilov, T., Šajn, R., (2015b). Litogenic distribution of rare earth elements in automorphic and alluvial soils from Bregalnica river basin. *Macedonian Journal of Chemistry and Chemical Engineering*, 34(1), 201–212.

Balabanova, B., Stafilov, T., Šajn, R., Tănăselia, C., (2016). Geochemical hunting of lithogenic and anthropogenic impacts on polymetallic distribution (Bregalnica River Basin, Republic of Macedonia). *Journal of Environmental Science and Health, Part A*, 51(13), 1180-1194 (2016).

Barandovski, L., Cekova, M., Frontasyeva, M.V., Pavlov, S.S., Stafilov, T., Steinnes, E., Urumov, V., (2008). Atmospheric deposition of trace element pollutants in Macedonia studied by the moss biomonitoring technique. *Environmental Monitoring and Assessment*, 138, 107–118.

Barandovski, L., Frontasyeva, M.V., Stafilov, T., Šajn, R., Pavlov, S., Enimiteva, V., (2012). Trends of atmospheric deposition of trace elements in Macedonia studied by the moss biomonitoring technique. *Journal of Environmental Science and Health, Part A*, 47(13), 2000–2015.

Barandovski, L., Stafilov, T., Šajn, R., Frontasyeva, M.V., Bačeva, K., (2013). Air pollution study in Macedonia by using moss biomonitoring technique. ICP-AES and AAS. *Macedonian Journal of Chemistry and Chemical Engineering*, 32(1), 89–107.

Barandovski, L., Frontasyeva, M.V., Stafilov, T., Šajn, R., Ostrovnaya, T., (2015). Atmospheric deposition of trace elements in Macedonia studied by the moss biomonitoring technique. *Environmental Science and Pollution Research*, 20(22), 16077–16097.

Boev, B., Stafilov, T., Bačeva, T., Šorša, A., Boev, I., (2013). The nickel smelter plant influence on the mineralogical composition of attic dust in Tikveš Valley, Republic of Macedonia. *Environmental Science and Pollution Research*, 20, 3781-3788.

Bowen H.T.M., (1979). *Environmental Chemistry of the Elements*. Academic Press, London.

Darnley, A. G., Björklund, A., Bolviken, B., Gustavsson, N., Koval, P.V., J. Plant, A., Steenfelt, A., Tauchid, M., Xuejing, X., Garrett, R.G., Hall, G.E.M., (1995). *A Global Geochemical Database for Environmental and Resource Management. Recommendations for International Geochemical Mapping*. Final report of IGCP project 259, UNESCO Publishing, Paris, pp. 122.

Davidson, C.M., Urquhart, G.J., Ajmone-Marsan, F., Biasioli, M., Duarte, A.D., Diaz-Barrientos, E., Grcman, H., Hossack, L., Hursthouse, A.S., Madrid, L., Rodrigues, S., Zupan, M., (2006). Fractionation of potentially toxic elements in urban soils from five European cities by means of a harmonised sequential extraction procedure. *Analytica Chimica Acta*, 565, 63–72.

D'Emilio, M., Caggiano, R., Macchiato, M., Ragosta, M., Sabia, S., (2013). Soil heavy metal contamination in an industrial area: analysis of the data collected during a decade. *Environmental Monitoring and Assessment*, 185(7), 5951–5964.

Dimovska, B., Šajn, R., Stafilov, T., Bačeva, K., Tănăselia, C. (2014a). Determination of atmospheric pollution around the thermoelectric power plant using a moss biomonitoring. *Air Quality, Atmosphere and Health*, 7, 541-557.

Dimovska, B., Stafilov, T., Šajn, R., Bačeva, K., (2014b). Distribution of lead and zinc in soil over the Bitola Region, Republic of Macedonia, *Geologica Macedonica*, 28, 87-91.

ELEM, (2007). An application for obtaining a permit with adjustment plan. *Power Plants of Macedonia*, Podružnica Bitola, Bitola.

Escarré, J., Lefèbvre, C., Raboyeau, S., Dossantos, A., Gruber, W., Cleyet Marel, J. C., Frérot, H., Noret, N., Mahieu, S., Collin, C., van Oort, F., (2011). Heavy metal concentration survey in soils and plants of the Les Malines mining district (Southern France): Implications for soil restoration. *Water, Air, & Soil Pollution*, 216(1–4), 485–504.

Ilić Popov, S., Stafilov, T., Šajn, R., Tánáselia, C., Bačeva, K., (2014). Applying of factor analyses for determination of trace elements distribution in water from river Vardar and its tributaries, Macedonia/ Greece. *The Scientific World Journal*, 2014 (2014), Article ID 809253, 1–11.

Ilić Popov, S., Stafilov, T., Šajn, R., Tánáselia, C., (2015). Distribution of trace elements in sediment and soil from river Vardar basin, Macedonia/Greece. *Journal of Environmental Science and Health, Part A*, 51(1), 1–14.

ISO 14869–1:2001, (2001). *Soil quality – Dissolution for the Determination of Total Element Content* – Part 1: *Dissolution with Hydrofluoric and Perchloric Acids*. International Organization for Standardization, Geneva, Switzerland.

Kabata-Pendias, A., Pendias, H. (2001). *Trace Elements in Soil and Plants*. Third Edition, CRC Press, Boca Raton.

Kabata-Pendias, A., Mukherjee, A.B. (2007). *Trace Elements from Soil to Human*. Springer, Heidelberg.

Kasassi, A., Rakimbei, P., Karagiannidis, A., Zabaniotou, A., Tsiouvaras, K., Nastis, A. K. Tzafeiropoulou, K., (2008). Soil contamination by heavy metals: Measurements from a closed unlined landfill. *Bioresource Technology*, 99(18), 8578–8584.

Kaymaz, I., (2005). Application of kriging method to structural reliability problems. *Structural Safety*, 27, 133–151.

Lazarevski, A., (1993). *Climate in Macedonia*. Kultura, Skopje. (In Macedonian).

Li, Z., Ma, Z., Van der Kuijp, T. J., Yuan, Z., Huang, L., (2014). A review of soil heavy metal pollution from mines in China: Pollution and health risk assessment. *Science of the Total Environment*, 468–469, 843–853.

Mirsal, I.A., (2008). *Soil Pollution: Origin, Monitoring & Remediation*. Springer Verlag, Berlin, Heidelberg.

Mico, C., Reatala, L., Peris, M., Sanchez, J., (2006). Assessing heavy metal sources in agricultural soils of an European Mediterranean area by multivariate analysis. *Chemosphere*, 65, 863–872.

Modis, K., Vatalis, K.I., (2014). Assessing the risk of soil pollution around an industrialized mining region using a geostatistical approach. *Soil and Sediment Contamination: An International Journal*, 23(1), 63–75.

Mustafa G., Komatsu S., (2016). Toxicity of heavy metals and metal-containing nanoparticles on plants. *Biochimica et Biophysica Acta (BBA) - Proteins and Proteomics*, 1864(8), 932-944.

Okedeyi, O.O., Dube, S., Awofolu, O.R., Nindi, M.M., (2014). Assessing the enrichment of heavy metals in surface soil and plant (*Digitaria eriantha*) around coal-fired power plants in South Africa. *Environmental Science and Pollution Research*, 21(6), 4686–4696.

Pavlov, K.., Bačeva, K., Stafilov, T., Vasileski, D., Toševska, S., (2012). Assessment of toxic metal pollution in some rivers in the Tikveš Basin, Republic of Macedonia. *International Journal of Environment Protection (IJEP)*, 2(12), 9-16.

Reimann, C., Äyräs, M., Chekushin, V.A., Bogatyrev, I.V., Boyd, R., Caritat, P. de, Dutter, R., Finne, T.E., Halleraker, J.H., Jæger, Ø., Kashulina, G., Lehto, O., Niskavaar, H., Pavlov, V.A., Räisänen, M.L., Strand, T., Volden, T., (2011). *Environmental Geochemical Atlas of the Central Barents Region*. Geological Survey of Finland and Geological Survey of Norway, Trondheim, ISBN: 978–3–510–65263–1.

Reimann, C., Birke, M., Demetriades, A., Filzmoser, P., O'Connor, P. (Eds.), (2014a). *Chemistry of Europe's Agricultural Soils – Part A: Methodology and Interpretation of the GEMAS Data Set*. Geologisches Jahrbuch (Reihe B), Heft 102, Schweizerbarth, Hannover, 528 pp.

Reimann, C., Birke, M., Demetriades, A., Filzmoser, P., O'Connor, P. (Eds.), (2014b). Chemistry of Europe's Agricultural Soils – Part B: General Background Information and Further Analysis of the GEMAS

Data Set. *Geological Yearbook* (Reihe B), Heft 103, Schweizerbarth, Hannover, 352 pp.

Šajn, R., Aliu, M., Stafilov, T., Alijagić, J., (2013). Heavy metal contamination of topsoil around a lead and zinc smelter in Kosovska Mitrovica/Mitrovicë, Kosovo/Kosovë. *Journal of Geochemical Exploration*, 134, 1–16.

Sakata, S., Ashida, F., Zako, M., (2003). Structural optimization using kriging approximation. *Computer Methods in Applied Mechanics and Engineering*, 192, 923–939.

Salminen, R., Batista, M.J., Bidovec, M., Demetriades, A., De Vivo, B., De Vos, W., Duris, M., Gilucis, A., Gregorauskiene, V., Halamic, J., Heitzmann, P., Jordan, G., Klaver, G., Klein, P., Lis, J., Locutura, J., Marsina, K., Mazreku, A., O'Connor, P.J., Olsson, S.Å., Rottesen, T., Petersell, V., Plant, J.A., Reeder, S., Salpeteur, I., Sandström, H., Siewers, U., Steenfelt, A., Tarvainen, T., (2005). *Geochemical Atlas of Europe,* Part 1, *Background Information, Methodology and Maps*. Geological Survey of Finland, Espoo.

Serafimovska, J. M., Arpadjan, S., Stafilov, T., Tsekova, K., (2013). Study of the antimony species distribution in industrially contaminated soils. *Journal of Soils and Sediments*, 13, 294–303.

Shahid M., Dumat C., Khalid S., Schreck E., Xiong T., Niazi N. K., (2017). Foliar heavy metal uptake, toxicity and detoxification in plants: A comparison of foliar and root metal uptake. *Journal of Hazardous Materials*, 325, 36-58.

Stafilov, T., Levkov, Z., (2007). *Summary of Vardar River Basin Field Survey*. European Agency for Reconstruction; Ministry of Environment & Physical Planning, Republic of Macedonia, Skopje.

Stafilov, T., Šajn, R., Pančevski, Z., Boev, B., Frontasyeva, M.V., Strelkova, L.P., (2008a). *Geochemical Atlas of Veles and the Environs*. Faculty of Natural Sciences and Mathematics, Skopje.

Stafilov, T., Šajn, R., Boev, Cvetković, J., Mukaetov, D., Andreevski, M., (2008b). *Geochemical Atlas of Kavadarci and the Environs*. Faculty of Natural Sciences and Mathematics, Skopje.

Stafilov, T., Peeva, L., Nikov, B., De Koning, A., (2009). Industrial hazardous waste in the Republic of Macedonia. In: *Applied Environmental Geochemistry – Anthropogenic İmpact on Human Environment in the SE Europe*, Ljubljana, *Proceedings Book* (R. Šajn, G. Žilbert, J. Alijagić, Eds.), pp. 108–112.

Stafilov, T., Aliu, M., Šajn, R., (2010a). Arsenic in surface soils affected by mining and metallurgical processing in K. Mitrovica region, Kosovo. *International Journal of Environmental Research and Public Health*, 7, 4050–4061.

Stafilov, T., Šajn, R., Pančevski, Z., Boev, B., Frontasyeva, M.V., Strelkova, L.P., (2010b). Heavy metal contamination of surface soils around a lead and zinc smelter in the Republic of Macedonia. *Journal of Hazardous Materials*, 175, 896–914.

Stafilov, T., Šajn, R., Boev, B., Cvetković, J., Mukaetov, D., Andreevski, M., Lepitkova, S., (2010c). Distribution of some elements in surface soil over the Kavadarci region. Republic of Macedonia. *Environmental Earth Sciences*, 61(7), 1515–1530.

Stafilov, T., Balabanova, B., Šajn, R., Bačeva, K., Boev, B., (2010d). *Geochemical Atlas of Radoviš and the Environs and the Distribution of Heavy Metals in the Air*. Faculty of Natural Sciences and Mathematics, Skopje.

Stafilov, T., Šajn, R., Sulejmani, F., Bačeva, K., (2011). *Geochemical Atlas of Kičevo and the Environs*. Faculty of Natural Sciences and Mathematics, Skopje.

Stafilov, T., Šajn, R., Balabanova, B., Bačeva, K., (2012). Distribution of heavy metals in attic and deposited dust in the vicinity of copper ore processing and ferronickel smelter plants in the Republic of Macedonia. In: *Dust: Sources, Environmental Concerns and Control*, L. B. Wouters, M. Pauwels (Eds.), Nova Science Publishers, Inc., Hauppauge, NY, pp. 57–98.

Stafilov, T., Šajn, R., Alijagić, J., (2013). Distribution of arsenic, antimony and thallium in soil in Kavadarci and its environs, Republic of Macedonia. *Soil and Sediment Contamination: An International Journal*, 22(1), 105–118.

Stafilov, T., (2014). Environmental pollution with heavy metals in the Republic of Macedonia. *Contributions, Section of Natural, Mathematical and Biotechnical Sciences, MASA*, 35(2), 81–119.

Stafilov, T., Balabanova, B., Šajn, R., (2014b). *Geochemical Atlas of the Region of the Bregalnica River Basin*. Faculty of Natural Sciences and Mathematics, Skopje.

Stafilov, T., Šajn, R., Sulejmani, F., Bačeva, K., (2014c). Geochemical properties of topsoil around the open coal mine and Oslomej thermoelectric power plant, R. Macedonia. *Geologia Croatica*, 67(1), 33–44.

Stafilov, T., Balabanova, B., Šajn, R., Rokavec, D., (2015). Variability assessment for lithogenic and anthropogenic distribution of trace and macroelements in water, sediment and soil samples. Case study: Bregalnica River Basin, Republic of Macedonia. In: *Advances in Environmental Research*, J. A. Daniels (Ed.), Nova Science Publishers, Inc., Hauppauge, NY.

Stafilov, T., Šajn, R., (2016). *Geochemical atlas of the Republic of Macedonia*. Faculty of Natural Sciences and Mathematics, Skopje.

Žibret G., Šajn R., (2010). Hunting for geochemical associations of elements: factor analysis and self-organizing maps. *Mathematical Geosciences*, 42, 681–703.

In: Soil Contamination
Editor: Jesper E. Lund

ISBN: 978-1-53613-266-3

Chapter 2

MONITORING THE POLYMETALLIC GEOCHEMISTRY OF SOIL ALONG THE KRIVA REKA RIVER BASIN DUE TO THE ANTHROPOGENIC IMPACT OF THE PB-ZN TORANICA MINE

Trajče Stafilov[1,*], Biljana Balabanova[2], Robert Šajn[3] and Svetlana Angelovska[1]

[1]Institute of Chemistry, Faculty of Science,
Ss. Cyril and Methodius University, Macedonia
[2]Faculty of Agriculture, Goce Delčev University, Macedonia
[3]Geological Survey of Slovenia, Ljubljana, Slovenia

[*] Corresponding Author Email: trajcest@pmf.ukim.mk.

ABSTRACT

Anthropogenic activities related to the exploitation of natural resources, technological processing and waste management represent a global problem of environment pollution. Previous studies have shown that certain areas in the territory of the Republic of Macedonia are affected by the anthropogenic introduction of high levels of different chemical elements into the soil. The area in the vicinity of the Toranica lead-zinc mine near the town of Kriva Palanka, eastern Macedonia, was examined regarding the lithogenic and anthropogenic distribution of 17 elements (Al, As, Ba, Ca, Cr, Cu, Fe, K, Li, Mg, Mn, Na, Ni, Pb, Sr, V and Zn). Alluvial soil samples (sediments and fluvisol) and automorphic soil samples (from the topsoil) were collected from the Kriva Reka River basin. The elemental contents were determined by atomic emission spectrometry with inductively coupled plasma (ICP-AES). Data processing was applied with combinations of multivariate statistical methods (factor analysis, principal component analysis and cluster analysis). The summary data for automorphic soil (topsoil and subsoil samples) reveal strong correlations between the elemental distribution in topsoil samples vs. subsoil samples along the Kriva Reka River. Anthropogenic impacts were observed in Zone 1 (the area affected by hydrothermal Pb-Zn exploitation). The soil surface layer was affected by enrichment with As, Mn, Pb and Zn (220, 3400, 1700 and 1100 mg/kg, respectively). Regression analysis revealed significant correlations for Cr, Ni, Pb and Zn in topsoil vs. subsoil (r: 0.92, 0.90, 0.91 and 0.95, respectively).

Keywords: chemical elements, toxic metals, monitoring, soil, environmental pollution, Kriva Reka River basin, Republic of Macedonia

1. INTRODUCTION

The emissions of metals into the environment occurs *via* a wide range of processes and pathways, including into the air (e.g., during combustion, extraction and processing), to surface waters (*via* runoff and releases from storage and transport) and to the soil (and hence into groundwater and crops) (Athar and Vohora, 1995; Dudka and Adriano, 1997; Agarwal et al., 2009; Acton 2013). The distribution of certain chemical elements, which at high levels represent a hazard to the environment, causes certain unwanted consequences to human health. Pollution of the environment with toxic metals has been topic of numerous studies focusing on industrial areas, areas where the exploitation and processing of natural resources (oil, ore, etc.) take place, or highly populated areas where traffic and communal waste represent the main sources of metals (Alloway and Ayres, 1997; Siegel, 2002; Järup, 2003).

The exploitation of mineral resources by humans requires the excavation, separation, transportation and dispersion of metals contained in fine dust (micro-particles). In this way, metals are introduced into the environment in much greater contents than those normally found in nature (Sengupta, 1993; Salomons, 1995). With the passage of time and long-term human activities, the contents of certain metals have been completely and permanently changed in relation to their natural existence in the environment (Wang et al., 2009). These changes can have a significant influence on the physiology and ecology of the organisms that adapt to survive under conditions with a higher metal contents described by VanLoon and Duffy (2000) as the *specific environment*. The distribution of different chemical elements, including the potentially toxic metals, creates specific conditions for living organisms (Krebs et al., 2006). Taking into account that metal contents in the environment are variable, it is important to identify regions with changing metal contents, differing from the natural distribution of elements in different segments of the biosphere (Athar and Vohora, 1995; Siegel, 2002; Artiola et al., 2004).

Anthropogenic activities related to the exploitation of natural resources, technological processes and waste management represent a global problem environmental pollution. The Republic of Macedonia does not diverge from this global framework of pollution of the environment with certain toxic metals. Previous studies have shown that certain areas in the territory of the Republic of Macedonia are affected by the anthropogenic introduction of high levels of different chemical elements into the soil (Stafilov et al., 2003, 2004, 2008a, 2008b, 2010c, 2010a, 2010b, 2011; Balabanova et al. 2010, 2011, 2012, 2013, 2014a, 2014b;

2016a, 2016b, 2016c, Angelovska et al., 2016; Barandovski et al., 2013; Stafilov and Šajn, 2016).

Macedonia was an important producer of metals in the former Yugoslavia, with significant contributions to the output of copper, steel, ferroalloys, lead, zinc, cadmium, silver and gold. One of the potential emission sources of hazardous heavy metals in air in the north-eastern part of the Republic of Macedonia is the Toranica lead-zinc mine, in the vicinity of the town of Kriva Palanka. This Pb-Zn mineral deposit has been exploited since 1987. Pb-Zn production from the mine occurred between 1987 and 2000, reaching total amount of ~3 million tons of excavated ore with an average contents of 6.5% Pb-Zn (Serafimovski and Aleksandrov, 1995; Serafimovski et al., 2006, 2007). Milling and flotation occurred at the mine and there is a tailings dam below the mine site with a culvert directed toward the Toranica River beneath the dam. The flotation plant for the production of Pb and Zn concentrates is situated close to the mine. Flotation tailing disposal leads to the uncontrolled introduction of large amounts from dust that contains a high content of Pb and Zn. Every process of exploitation of mineral resources leads to changing the natural distribution regarding certain chemical elements in environment.

The exposure of mining and flotation tailings, as well as the exposure of the open ore pit to the air streams, leads to distribution of the finest dust (Sengupta, 1993; Athar and Vohora, 1995). Additionally, rainwater washes the ore tailings, leading to water extraction of the available metals. In this way, metals and other potentially toxic chemical elements are introduced into the soil and water (surface and underground). At the same time, particles from the flotation and mining tailings are continually distributed by the wind. These particles, depending on the weather conditions, can be distributed in the air for long or short periods of time and thus be deposited shorter or greater distances from the emission source (Van het Bolcher et al., 2006).

The most suitable technique for this type of study is the application of relevant monitoring that will enable a more lasting and efficient determination of environmental quality in the investigated area (Thornton, 1996; Artiola et al., 2004). Monitoring programs were implemented as a component of numerous analytical research activities many years ago, and in recent decades, their application has been highly promoted both at the regional and global level (Rühling and Tyler, 1971, 1973; Koljonen, 1992; Sengupta, 1993; Hoenig, 2001). This is due to the fact that monitoring programs do not require the use of expensive technical equipment, and the analytical results reflect the situation over a long period of time.

2. Area Under Investigation

2.1. Geographic Characterization of the Investigated Area

The study area is the region of the Kriva Reka River basin. A total area of 1250 km^2 was monitored, limited by the coordinates N: 42°10' - 42°20' and E: 22°17' – 22°28', located in the north-eastern part of the Republic of Macedonia (Figure 1). The Toranica Pb-Zn mine is located in the eastern part of the study area.

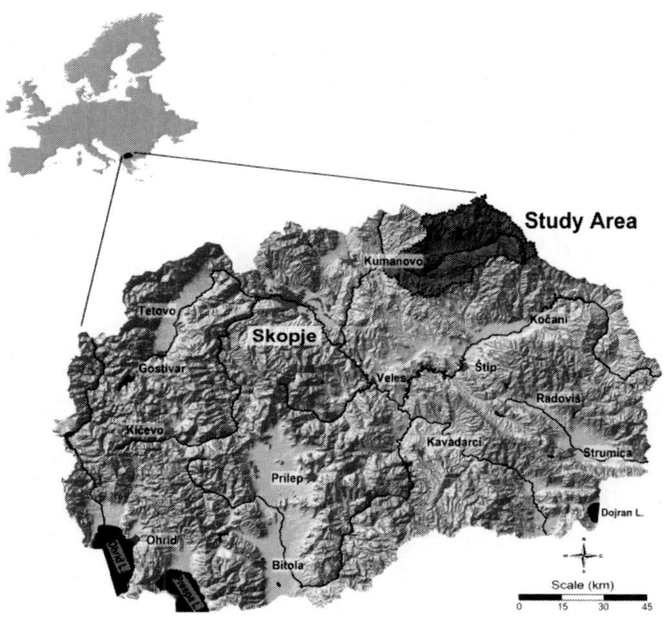

Figure 1. The investigated area on the territory of the Republic of Macedonia.

The study area is characterized by a moderate continental climate. The altitude varies between 600 and 1500 m. Most frequent winds in the region are those from the west with frequency of 199 ‰ and 2.7 m/s speed, and winds from the east with frequency of 124 ‰ and speed of 2.0 m/s (Lazarevski, 1993). Climatic conditions in the region allows for the air distribution of fine dust particles generated as a result of exposed flotation tailings. Taking into consideration the diverse morphological-geographical structure of the investigated area, the climate conditions also vary in the different sub-regions.

The average annual precipitation amounts to about 500 mm with significant variations from year to year, as well as in the different sub-regions (Lazarevski, 1993). The precipitation is mostly related to and conditioned by Mediterranean cyclones (Lazarevski, 1993). During the summer period, the region is most often found in the center of the subtropical anticyclone, which causes warm and dry summers. From the central area of the region, as the driest area, the average annual precipitation increases in all directions, because of an increase either in the influence of the Mediterranean climate or in altitude (Lazarevski, 1993). The precipitation information refers to the annual average precipitation as measured by meteorology stations in mm, with information provided by the Hydro-meteorological Administration. Regarding the annual total number of sunny hours, there are about 6.0 hours per day in this area. The region is comprised of about ten climatic-vegetation soil areas with highly heterogeneous climate, soil and vegetation characteristics (Lazarevski, 1993).

2.2. Generalized Geology of the Investigated Area

The region of the investigated area geologically is composed of several sub-regions (zones). A generalized geology map is provided in Figure 2. The study area is a part of the Osogovo volcanic and ore-bearing district (Serafimovski et al., 2006a). The volcanic rocks in the investigated area occur as elongated dykes with the northwest direction (Serafimovski et al.,

2006a, 2006b). Larger areas are covered by various rocks of different origin. Quartz graphite schists are of special importance, because they are one of the most important host rocks for the Pb-Zn mineralization (Boev and Yanev, 2001). They are usually incorporated with cipolin marbles, which are also very favorable for metasomatic Pb-Zn mineralization (Serafimovski et al., 2006a, 2006b). The disjunctive fault structures are the most abundant in the Toranica ore deposit.

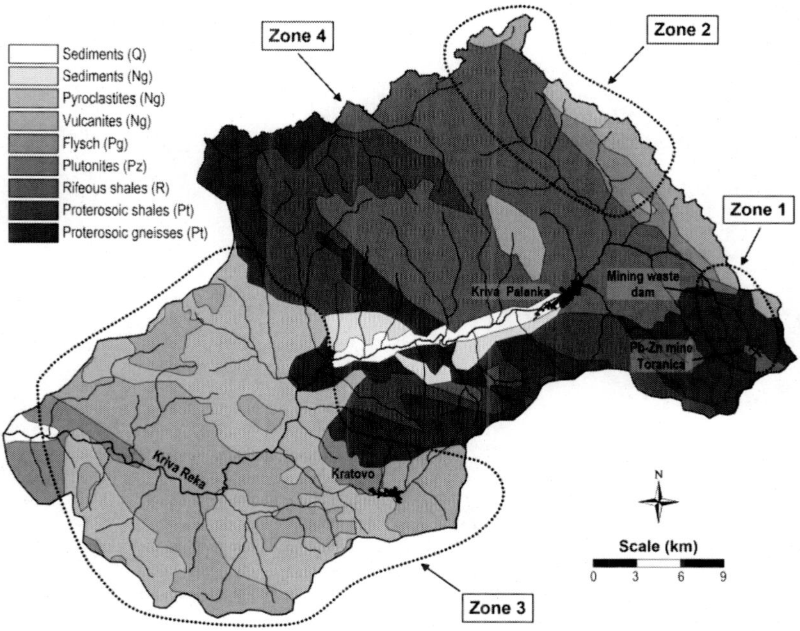

Figure 2. Lithological map of study area with defined zones (simplified according to Geological map of SFR Yugoslavia, Rakićević et al., 1968).

According to the constructed lithological map presented in Figure 2, the investigated area is divided into five zones. Zone 1 covers the area of where the Toranica lead and zinc mine is located. In this zone the dominant geological occurrences are Proterozoic gneisses. The area around the Toranica mine contains shale, while in the northern part of this area contains plutonium volcanic and marl. Zone 2 represents an urban area and contains shale sediments in the geological structure of this area. Zone 3 is in the north-eastern part of the examined area. Anthropogenic influences are found in two zones (Zone 1 and 2). Zone 1 is the sub-area where the Toranica mine is located, while the Zone 2 is characterized as an urban area, i.e., where the town of Kriva Palanka is located. The geological structure of this area in the central and western part includes shale, while in the rest of the zone occurs pyroclastite and vulcanite occur. The central part of the examined area is designated as Zone 4 and is characterized by the occurrence of Proterozoic shales and gneisses and Rifeous shales, Proterozoic gneiss and schist which are prevalent in the northern and western part of this area. The area downstream of the Kriva Reka River is characterized by the presence of sedimentary rocks. The detailed geological investigation revealed that ore mineralization is closely related to Neogene volcanism. The ore mineralization is concentrated in the Zone 5 and partially in Zone 3 (Figure 2).

3. Materials and Methods

3.1. Samples Analysis

The monitoring activities included determining the contents of 17 elements (Al, As, Ba, Ca, Cr, Cu, Fe, K, Li, Mg, Mn, Na, Ni, Pb, Sr, V and Zn) in soil samples from the Kriva Reka River basin. For each location where samples were taken, the location characteristics were recorded (geographic coordinates and altitude) using a global positioning system. This positioning, that is, specifying the position of the locations from where samples were taken, was required for the construction of maps for the distribution and deposition of the analyzed elements in the investigated area.

3.1.1. Sampling Network of Soil Samples

The location network had a density of 5 × 5 km, in order to enable mapping and specifying of the distribution of the chemical elements examined (Figure 3). At each location a sample of topsoil was taken at a depth of 0-5 cm, as well as a sample of subsoil, taken at a depth of 20-30 cm. The collection of soil samples was undertaken in accordance with the relevant standards for the collection of this type of sample from the environment (Reimann et al., 2012). The locations for the collection of soil samples were previously determined by constructing a network for sample collection. Depending on the conditions of the site and its accessibility, the samples were collected in the surroundings of the specified locations, while recording the relevant coordinates. At every location, five topsoil and subsoil samples were collected. At the specified location, a square with an area of 10 × 10 m was marked of. Four samples were collected from the angles of the square while the fifth sub-sample was collected from the intersection of the square diagonals. In this way, a representative composite sample of topsoil was collected. From the holes formed by taking of the topsoil sub-samples, the sub-soil sub-samples were also collected, for the formation of a representative sample of a sub-soil sample. Randomly were also collected alluvial soil samples along the course of the river Kriva Reka.

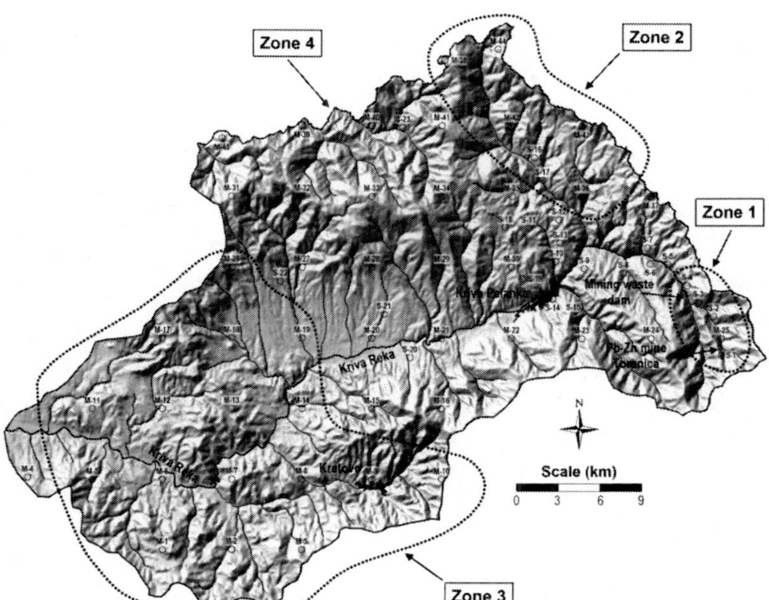

Figure 3. Locations for collection of soil samples in the investigated area.

3.2. Analytical Reagents

All used reagents and standards were of the analytical purity level: nitric acid, HNO_3, 69%, ultra pure (Merck, Germany), hydrogen peroxide, H_2O_2, 30% p.a. (Merck, Germany), concentrated hydrochloric acid, HCl 37%, p.a. (Merck, Germany), concentrated hydrofluoric acid, HF, p.a. (Merck, Germany) and concentrated perchloric acid, $HClO_4$, p.a. (Alkaloid Skopje). Redistilled water was used for the preparation of all solutions. The standard solutions for the elements examined were prepared by dissolution of the basic multielement standard solution with concentration of 1000 mg L^{-1} (11355-ICP, Multi Element Standard). A series of standard solutions were prepared in a linear range in several concentration areas as, *series 1*: 1, 2, 3, 5, 10 μg/L and *series 2*: 10, 20, 30, 50, 100 μg/L for elements in traces, for the remaining elements; *series 3*: 0,1; 0,2; 0,3; 0,5; 1 mg/L; *series 4*: 1, 2, 3, 5, 10 mg/L and *series 5*: 10, 20, 30, 50, 100 mg/L for the elements in macro contents.

3.3. Sample Preparation

For the digestion of soil and sediment samples open digestion method was applied with a mixture of concentrated mineral acids. The soil and sediment samples brought into the laboratory were subjected to cleaning and homogenization, and then were dried at room temperature, or in a drying room at 40°C, to a constantly dry mass. Then, the samples were passed through a 2 mm sieve and finally were homogenized by grinding in a porcelain mortar until reaching a final size particle of 25 μm. Following physical preparation, the samples were chemically prepared by wet digestion, applying a mixture of acids in accordance with international standards (ISO 14869-1:2001). In this way, the digested soil and sediment samples were prepared for determining the contents of the different elements using atomic emission and mass spectrometry.

3.4. Instruments Used For Determining the Content of Elemental in Various Samples

Atomic emission spectrometry with inductively coupled plasma (ICP-AES) was used for determination of 17 elements contents. The optimization of the instrumental techniques was performed are given as follows: optimal parameters for the RF generator were an operating frequency of 40.68 MHz, power output of 1500 W; the power output stability was better than 0.1%. Spectrometer optimization was performed according to the manufacturer's recommendations. The settings of the program were: plasma Ar flow rate 15 L/min, auxiliary Ar flow rate 1.5 L/min, nebulizer Ar flow rate 0.75 L/min, pump speed 25 rpm, stabilization time 30 s, rinse time 30 s, sample delay 30 s, with three replicates for the quantification measurements. The sensitivity defined as the lower limit of detection was assessed and shown in Table 1.

The QC/QA of the applied techniques was performed by the standard addition method, and it was found that the recovery for the elements investigated by ICP-AES ranged from 90.8% to 109.2%.

Table 1. Limit of detection (in mg/kg) and wavelength (WL) for the analyzed elements for ICP-AES

Element	WL, nm	LOD	Element	WL, nm	LOD
As	188.9	0.01	Mg	279.5	0.025
Al	396.1	0.0125	Mn	257.6	0.0015
Ba	455.4	0.025	Na	589.5	2.5
Ca	370.6	0.025	Ni	231.6	0.25
Cr	267.7	0.05	Pb	220.3	0.1
Cu	324.7	0.0125	Sr	407.7	0.025
Fe	238.2	0.006	V	292.4	0.05
K	766.5	5.0	Zn	213.8	0.003
Li	670.7	0.05			

3.5. Data Processing

3.5.1. Basic Statistical Analysis and Transformation

All data on the content of the investigated elements were statistically processed using statistical software (Stat Soft, 11.0) to perform parametric and non-parametric analysis. A basic descriptive statistical analysis was conducted on the values of the element contents in all types of samples. At the same time, normalization tests were performed and, based on the results and the visual check of the distribution histograms, the distribution of the data as independent variables (element contents) was determined.

Data transformation means performing the same mathematical operations on each component of the original data. If the original data are simply multiplied or divided by a specific quotient or they are continually subtracted or added, this is linear transformations; which does not change the data form (that is, the distribution) and accordingly does not help in the normalizing the data distribution. This is most often a characteristic of data on variables from the environment whose distributions are log-normal or positively curved. Therefore, it is necessary to first subject the data to transformation, that is, to perform normalize the distribution. Logarithmic transformation is widely applied in order to normalize the positively skewed data distributions. The aim of the application of data transformation is to actually decrease the difference between extreme values. For data normalization, the *Box-Cox* transformation method (Box and Cox, 1964) was applied.

3.5.2. Bivariate and Multivariate Analysis

The level of correlation of the values of the chemical element contents in all types of samples was estimated using bivariate statistics (using linear correlation of coefficients, *r*), with significance levels of $p < 0.05$ and $p < 0.01$. For a clearer overview, the correlation coefficients are represented in a matrix of correlation coefficients. In the case of a large number of variables (for example $n > 20$), bivariate analysis is not useful. Therefore, the application of multivariate analysis enables the reduction of the number of variables by creating new synthetic variables, again dependent on the correlation (*r*).

For the purposes of this study, two methods of multivariate analysis were used, i.e., factor analysis (FA) and cluster analysis (CA). The primary aim of factor analysis is to explain the variation in a set of multidimensional data using as few factors as possible and to reveal the hidden data structure (Šajn 2009; Žibret and Šajn, 2010). In other words, the factor analysis is performed on a great number of variables in order to create a small number of new, synthetic variables, called factors. The factors contain a great deal of information about the original variables, and they may have some significance. Factor analysis is performed on variables that are standardized to zero values of a standard deviation unit. Data can be transformed and/or standardized.

Cluster analysis (CA) was used with a similar aim as factor analysis, in cases where high values for the loadings of certain variables were not acquired. In such cases, a data processing method is applied based on a matrix of remoteness (distances) or nearness (similarities). To establish the dependence among a series of variables in a multidimensional system, the Euclidean distance matrix is most commonly used.

3.5.3. Visual Representation of the Chemical Element Distribution

To construct the maps for the individual element distributions, as well as for the factor associations, the universal *Kriging* method was applied. Kriging is an optimal prediction method, intended for geophysical variables with a continuous data distribution. The obtained values of variables can sometimes be random, but their variance is not described by a geometric function. This method performs projection of an object by using the values of certain parameters that describe its position (latitude, longitude and oval height), i.e., any object on Earth can be spatially defined (Kaymaz, 2005). All data about variables are organized in several possible ways; the most commonly used is raster. Generally, raster consists of a matrix of cells (pixels), arranged in rows and columns, where each cell contains values that provide information about the variables. Digital aerial photography, satellite imagery, digital images and scanned maps can be used as a raster. The Kriging interpolation method provides the output of each raster cell by calculating the average load of nearby vectors. Kriging method analyzes the statistical variation of the values of different distances and at various positions to determine the shape and size of a specified point for examination as a set of load factors (Sakata et al., 2003).

To make the distribution maps the Kriging method was applied with linear variogram interpolation. As area limits, the percentile values of the distribution of the interpolated values were considered.

4. Results and Discussion

4.1. Basic Descriptive Statistics of the Selected Elements

The descriptive data statistics for the element contents in soil are shown in Table 2. Data normalization was done using the Box-Cox transformation, so further data processing was conducted using Box-Cox normalized values. The distribution of aluminum was characterized by a median value of 5.4% in topsoil (TS) and 5.3% in subsoil (SS). The range of this element distribution was from 1.3% to almost 10%. The median aluminum contents for Macedonia (6.6%) and Europe of (5.6%) are similar, showing no significant enrichment in this area (Salminen et al., 2005; Stafilov and Šajn, 2016). Arsenic in enriched content can be found as very toxic. According to the data obtained from the present study, the mean value for arsenic in topsoil was 27 mg/kg, which is almost three times higher than the Macedonian soil average, and two times higher than European topsoil mean content (Salminen et al., 2005; Stafilov and Šajn, 2016). Comparative analysis for the barium content also showed significant enrichment of this element compared with Macedonian and European data (Salminen et al., 2005; Stafilov and Šajn, 2016). The chromium mean content was lower than the corresponding value obtained for Macedonia (88 mg/kg). Similar to Cr, copper showed insignificant enrichment in the topsoil layer compared with the mean value for Cu for Macedonia; however, compared to the European topsoil content, an enrichment value of ~2 was obtained (Salminen et al., 2005). The copper content was 110 mg/kg in the subsoil layer, with a lower content in topsoil layer (97 mg/kg). However, the TS/SS ratio was almost 1, due to the lithogenic enrichment of this element in the investigated area. Lithium was characterized with a predominantly lithological distribution for the topsoil/subsoil layer, ranging from a minimum of 7.8 mg/kg (in subsoil layer) to a maximum of 78 mg/kg (subsoil layer). The Fe content did not differ significantly between soil layers. The maximum value for iron was found in the subsoil layer (9.8 mg/kg), indicating a characteristically lithogenic deposition. The mean value for Pb (94 mg/kg) compared to the same value for the whole territory of the Republic of Macedonia (34 mg/kg) showed an enrichment value of ~3 (Stafilov and Šajn, 2016). The lead contents in the whole investigated area ranged from 10 to 1700 mg/kg. High variability was obtained in the distribution of lead, which also confirms the differences in the values for the average and the median (52 mg/kg). Compared with some sub-areas in the Republic of Macedonia, such as the Kamenička River (Sasa mine) and the Zletovska River (Zletovo mine), where the median value is as high as 3100 mg/kg, the present data do not indicate significant pollution (Balabanova et al., 2017). Lead distribution is also related to the areas of Neogene pyroclastites.

The values obtained for the zinc contents in the soils ranged from 61 to 1400 mg/kg, with a mean of 160 mg/kg, which is almost two times higher than the mean value obtained for the Republic of Macedonia and 2.4 times higher than the European average (Salminen et al., 2005; Stafilov and Šajn, 2016). The distribution of this element was almost to that of lead probably due to the two major areas of Pb-Zn mineralization in the region. Higher contents of Pb and Zn were found in soil samples very close to the pollution source, despite the much lower median values for Pb and Zn (Table 2).

Lead-zinc deposits usually are characterized by high Mn contents as well (Balabanova et al., 2017). The Mn contents ranged from 400-3600 mg/kg, and can be considered as geo-enrichments in the study area. The characterization of the Cr and Ni contents showed intensive similarities.

Table 2. Descriptive statistics of measurement, n = 78/68

	Material	Unit	X	Md	Min	Max	S	S_X	CV	$A_{(N)}$	$E_{(N)}$	$A_{(B)}$	$E_{(B)}$
Al	Topsoil	%	5.4	5.4	1.4	8.3	1.9	0.23	35	-0.13	-0.98	-0.27	-1.02
Al	Subsoil		5.3	5.3	1.3	9.8	2.1	0.26	40	2.18	-0.96	-0.10	-0.96
As	Topsoil	mg/kg	27	23	2.7	220	30	3.7	110	0.94	25.0	-0.12	-0.56
As	Subsoil		33	26	2.1	360	52	6.4	155	1.81	27.0	0.07	0.11
Ba	Topsoil		530	420	130	1300	290	36	56	1.70	-0.05	-0.15	-0.66
Ba	Subsoil		560	450	120	2100	370	45	65	1.08	3.56	0.11	-0.41
Ca	Topsoil	%	1.5	1.2	0.10	6.1	1.1	0.14	75	0.90	3.35	-0.01	-0.07
Ca	Subsoil		1.4	1.2	0.11	6.0	1.1	0.14	79	5.76	2.73	-0.01	-0.57
Cr	Topsoil	mg/kg	84	68	11	320	65	8.0	78	6.09	3.14	-0.11	-0.41
Cr	Subsoil		84	66	12	380	72	8.9	86	-0.30	4.52	0.09	-0.49
Cu	Topsoil		32	28	13	97	13	1.6	42	4.88	8.13	-0.13	0.53
Cu	Subsoil		32	28	11	110	18	2.2	56	4.15	7.78	0.06	0.46
Fe	Topsoil	%	4.9	4.6	2.6	9.6	1.4	0.17	28	1.63	1.40	0.06	-0.02
Fe	Subsoil		4.9	4.8	2.3	9.8	1.6	0.19	32	0.83	0.91	-0.05	0.15
K	Topsoil		1.7	1.6	0.40	3.3	0.77	0.09	46	1.43	-0.86	-0.19	-0.86
K	Subsoil		1.7	1.7	0.55	4.7	0.88	0.11	50	1.59	0.81	0.10	-0.70
Li	Topsoil	mg/kg	22	18	8.0	73	12	1.5	56	1.96	5.85	0.01	-0.14
Li	Subsoil		22	19	7.8	78	12	1.5	56	1.69	6.10	0.03	-0.37
Mg	Topsoil	%	1.1	1.0	0.29	2.2	0.42	0.05	37	2.47	0.29	-0.18	0.71
Mg	Subsoil		1.1	1.0	0.35	2.2	0.45	0.06	41	0.95	0.37	0.16	-0.04
Mn	Topsoil	mg/kg	1400	1300	410	3400	580	71	41	0.93	1.12	0.03	0.02
Mn	Subsoil		1400	1300	400	3600	630	78	45	0.39	1.31	-0.03	0.17
Na	Topsoil	%	1.2	1.1	0.32	2.4	0.46	0.06	38	0.91	0.62	0.00	0.30
Na	Subsoil		1.2	1.1	0.32	2.7	0.48	0.06	41	2.19	1.33	0.03	0.53
Ni	Topsoil	mg/kg	36	31	2.5	260	35	4.4	97	2.11	25.0	0.04	1.14
Ni	Subsoil		37	28	2.7	250	37	4.5	100	0.85	18.2	0.00	0.62
Pb	Topsoil		94	52	10	1700	210	26	227	1.01	49.0	0.07	0.16
Pb	Subsoil		89	44	8.4	1700	220	26	242	0.80	54.7	-0.05	0.00
Sr	Topsoil		210	120	27	940	200	25	95	0.87	3.03	0.09	-0.57
Sr	Subsoil		210	120	21	840	200	25	98	0.77	2.07	-0.01	-0.54
V	Topsoil		140	130	57	340	57	7.1	41	3.60	1.41	0.22	-0.21
V	Subsoil		150	140	33	350	61	7.5	42	4.27	1.15	-0.18	0.31
Zn	Topsoil		160	140	64	1100	130	16	81	7.13	39.2	-0.04	0.99
Zn	Subsoil		170	140	61	1400	160	20	97	6.67	43.2	-0.08	0.49

X –mean; Md – median; Min – minimum; Max – maximum; S – standard deviation; S_X – standard error of mean; CV – coefficient of variation (%); A – skewness; E – kurtosis; (N) – normal values, (B) – Box-Cox transformation.

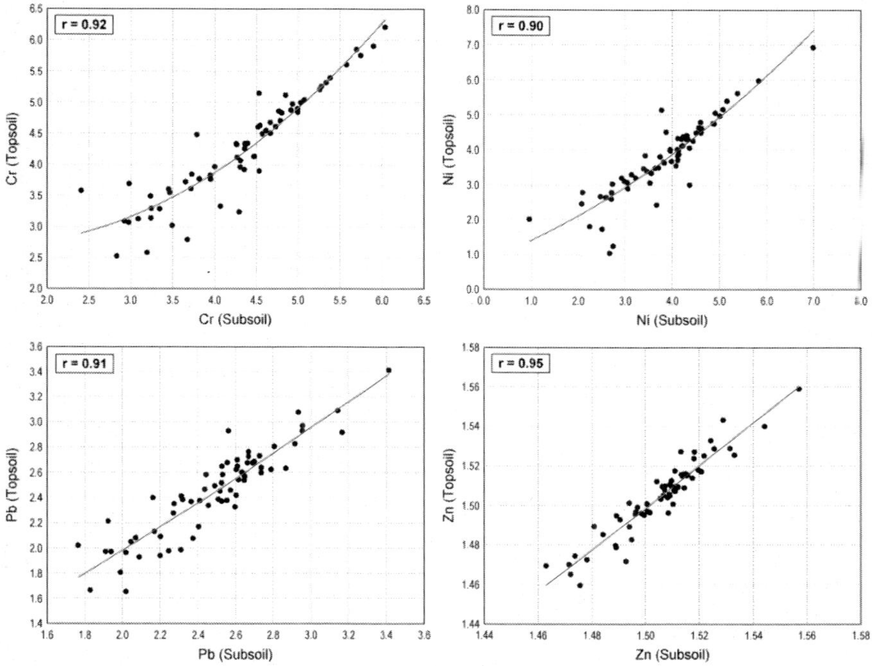

Figure 4. Regression plots of Cr, Ni, Pb and Zn (topsoil vs. subsoil), n = 78/68.

The minimum content of Cr and Ni was found in the topsoil layer and did not differ significantly from the content in the subsoil layer (Table 2).

The values for the concentration of the elements in the soil were assessed as a functional dependence of two variables. Regression plots are provided for the distribution of the elements in the topsoil vs. subsoil layer. A very high significant correlation ($r > 0.90$) was obtained for Cr, Ni, Pb and Zn (Figure 4).

Because of the complexity of the investigated area, the data on the metal contents in soil samples are given separately according to the zones (Table 3). Table 3 contains average data from the present analyses divided into zones. It can be seen that only the aluminum did not vary significantly between zones. For the rest of the analyzed elements, significant differences were obtained for the elements content (Table 3). When the element contents were compared according to the layers, no significant differences were obtained for all of the analyzed elements (F-ratio, Table 3). Only for lead were higher values obtained for the content in the topsoil layer vs. the subsoil layer, which indicates a significant interruption of the lithogenic distribution of this element in the investigated area. This occurrence was found in all four zones in the investigated area.

In addition to monitoring of the contents of individual chemical elements in soil, the correlation relationships among these elements were also analyzed. For this reason, a matrix of correlation coefficients was constructed (Table 4). The highest correlation coefficients were found for the following relationships: Ni-Cr (0.91), K-Ba (0.90), Fe-V (0.90), Pb-Zn (0.81), Pb-Ba; Sr-Ca; Zn-Mn (0.79), Fe-Mn (0.72), Pb-K and As-Zn (0.71).

Table 3. Averages of chemical elements and results of F-test regarding to soil layer and determined zones

	Unit	Zone 1		Zone 2		Zone 3		Zone 4		F ratio		F ratio	
		TS	SS	TS	SS	TS	SS	TS	SS	(Layer)		(Zones)	
n		12/5	12/5	7	7	17	17	41/38	41/38				
Al	%	4.7	4.7	6.9	6.1	6.0	5.5	4.9	5.2	1.28	NS	2.61	NS
As	mg/kg	52	70	64	67	28	31	17	18	1.20	NS	14.2	*
Ba		920	1100	330	310	780	800	380	410	1.12	NS	27.4	*
Ca	%	0.88	0.71	1.4	1.2	2.2	2.1	1.3	1.2	1.18	NS	10.1	*
Cr	mg/kg	68	75	160	180	38	33	92	90	1.09	NS	28.6	*
Cu		45	60	45	48	27	27	29	28	1.41	NS	11.1	*
Fe	%	4.7	4.9	7.5	7.6	4.6	4.6	4.5	4.7	1.27	NS	13.8	*
K		2.1	2.3	1.1	1.2	2.4	2.5	1.3	1.4	1.09	NS	23.6	*
Li	mg/kg	22	24	29	30	21	21	20	20	1.04	NS	5.02	*
Mg	%	1.1	1.1	1.8	1.9	0.84	0.83	1.1	1.1	1.08	NS	23.6	*
Mn	mg/kg	1900	2000	2000	2000	1400	1400	1200	1200	1.21	NS	11.8	*
Na	%	1.4	1.2	1.3	1.4	0.94	1.0	1.3	1.3	1.10	NS	6.37	*
Ni	mg/kg	32	40	61	66	18	14	36	37	1.20	NS	25.6	*
Pb		510	480	40	33	100	93	44	42	1.13	NS	17.6	*
Sr		89	79	120	93	450	440	130	130	1.16	NS	30.0	*
V		130	140	230	230	150	160	120	130	1.19	NS	12.9	*
Zn		410	490	170	180	160	150	130	130	1.27	NS	14.4	*

n – number of observation; F – ratio (ANOVA); NS – no significant; * – significance at $p < 0.05$; TS-top soil; SS-sub soil.

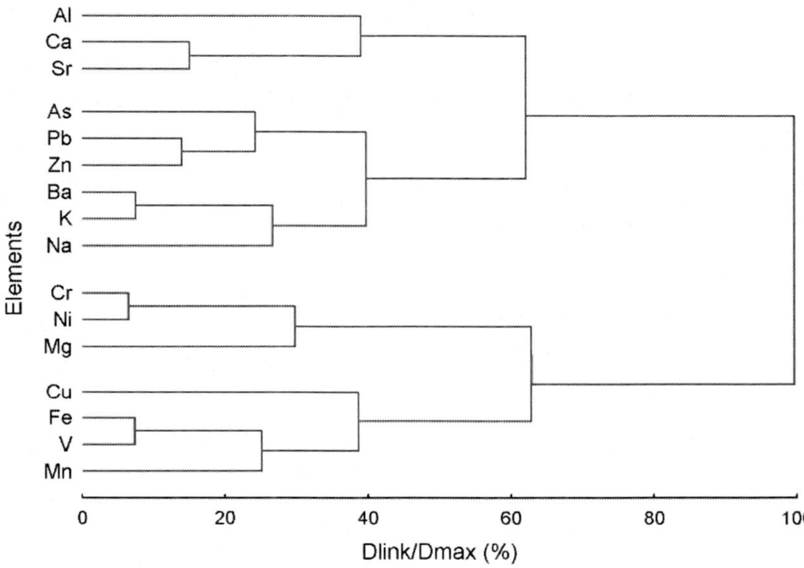

Figure 5. Cluster analysis dendrogram – element relationship (n = 78/68, 16 selected elements).

Table 4. Matrix of correlation coefficients, n = 78/68

Al	1.00															
As	0.28	1.00														
Ba	0.33	0.55	1.00													
Ca	0.46	0.21	0.19	1.00												
Cr	0.17	0.22	-0.27	-0.03	1.00											
Cu	0.22	0.50	0.17	0.10	0.55	1.00										
Fe	0.51	0.38	-0.05	0.36	0.36	0.50	1.00									
K	0.24	0.48	**0.90**	0.14	-0.32	0.10	-0.01	1.00								
Li	0.10	0.30	0.16	-0.26	0.32	0.32	0.21	0.28	1.00							
Mg	0.33	0.15	0.39	0.18	0.60	0.41	0.66	0.37	0.28	1.00						
Mn	0.52	0.62	0.40	0.29	0.33	0.62	**0.72**	0.35	0.43	0.44	1.00					
Na	-0.26	-0.45	-0.63	-0.22	-0.09	-0.25	-0.04	-0.68	-0.34	0.22	-0.39	1.00				
Ni	0.13	0.19	-0.23	-0.10	**0.91**	0.54	0.28	-0.25	0.40	0.59	0.30	-0.10	1.00			
Pb	0.30	0.66	**0.79**	0.27	-0.03	0.39	0.14	**0.71**	0.24	-0.14	0.59	-0.64	-0.04	1.00		
Sr	0.51	0.28	0.54	**0.79**	-0.20	0.05	0.13	0.47	-0.13	-0.17	0.26	-0.50	-0.22	0.48	1.00	
V	0.36	0.37	-0.02	0.50	0.24	0.46	**0.90**	0.01	0.14	0.49	0.65	-0.10	0.13	0.18	0.27	1.00	
Zn	0.36	**0.71**	0.61	0.15	0.23	0.61	0.52	0.57	0.37	0.19	**0.79**	-0.49	0.20	**0.81**	0.26	0.46	1.00
	Al	As	Ba	Ca	Cr	Cu	Fe	K	Li	Mg	Mn	Na	Ni	Pb	Sr	V	Zn

Values in range 0.5 - 0.7 (good association) are underlined and in range 0.7 - 1.0 (strong association) are bolded; Box-Cox transformed values used.

Table 5. Matrix of dominant rotated factor loadings – normal values (n = 78/68, 16 selected elements)

	F1	**F2**	**F3**	**F4**	**Comm**
Ba	**0.90**	-0.22	0.20	-0.08	90.3
Pb	**0.89**	-0.02	0.16	0.15	84.3
K	**0.86**	-0.27	0.15	-0.07	84.4
Zn	**0.78**	0.17	-0.02	**0.51**	89.8
Na	**-0.75**	-0.18	-0.33	0.16	72.7
As	**0.70**	0.18	0.04	0.38	66.7
Ni	-0.02	**0.96**	-0.06	0.10	94.0
Cr	-0.05	**0.94**	-0.02	0.19	92.7
Cu	0.39	**0.56**	-0.06	0.48	69.0
Ca	0.04	-0.08	**0.88**	0.27	84.6
Sr	0.38	-0.18	**0.86**	-0.00	92.1
Al	0.21	0.16	**0.63**	0.34	58.6
Fe	0.03	0.20	0.20	**0.92**	92.6
V	0.04	0.05	0.29	**0.88**	85.5
Mn	**0.53**	0.25	0.15	**0.70**	85.4
Mg	-0.29	**0.55**	0.08	**0.63**	78.7
Prp. Totl	29.5	17.6	14.2	21.4	82.6
Expl. Var	4.71	2.81	2.27	3.42	
Eigen Val	6.13	3.96	1.99	1.14	

F1- ... F4 – Factor loadings; Com – Communality in %; Prp. Totl – Principal total variance in %; Expl. Var – Explained variance; Eigen Val – Eigenvalues; Dominant values are bolded.

In order to obtain a better picture for the interdependence of the analyzed elements, multivariate factor analysis was conducted. Based on the constructed matrix of correlation coefficients, statistical screening for Eigenvalues (E-values) was performed in order to single out only the significant synthetic variables ($E > 1$). The values for the isolated factor loadings (F1, F2, F3, F4) are presented in Table 5.

Based on the performed multivariate processing of the values of the analyzed elements' contents, three dominant factors were singled out: factor 1 or F1, had a dominant loading value of 6.13 and explained 29.5% of the total variability (82.6%). This factor associates the following elements Pb-Zn-As-Ba-K-Na (Table 5). The highest loading value in this geochemical association of elements was obtained for barium (0.90). The second factor associates the elements Ni-Cr-Cu, with a variability of 17.6% and E-value of 3.96 (Table 5). The third distinguished factor is represented by the Ca-Sr-Al association, which covers 14.2% of the total variability of the factor loadings matrix, with a E-value of 1.99. The fourth elemental association (F4) intercorrelates Fe-V-Mn-Mg, with total variance of 21.4% and the lowest E-value (1.14).

Multivariate cluster analysis was also applied in order to determine the factor analysis significance and the stability of the new synthetic variables, that is, associations of elements. The dendrogram of the distances among the individual elements is presented in Figure 5.

4.2. Areal Distribution of the Elements (Geochemical Markers)

By visualization of nearness, that is, the remoteness of the individual elements, the dependence of a single association of elements can also be observed. This is particularly emphasized in relation to Factor 1: Pb-Zn-As-Ba-K-Na. The areal distribution of this factor association is shown in Figure 6. The arsenic distribution is specifically related to the old volcanism in the western region of the investigated area. Characteristic and potentially toxic contents of As in the soil were found in the Zone 1 (>38 mg kg^{-1}). The barium distribution is similarly to that of arsenic, is closely related to the volcanism in the investigated area. Specific for barium is the fact that its natural enrichment is related to both old and the young volcanism in the investigated area. Paleogenic volcanic sedimentary rocks represent young volcanism and are found distributed along the eastern border of the investigated area. This region is characterized with naturally distributed barium (2300 mg kg^{-1}). Neogene pyroclastites are the dominant geogenic formations that represent old volcanism in the investigated area. These Ba contents exceed the optimum and action values in accordance with the Dutch standards (200 and 625 mg/kg, respectively). The basic geogenic marker of this group is related to the distribution of lead and zinc. Both areas lie on metamorphic and volcanic rocks originating from the Tertiary. The Pb-Zn mineralization in this area characteristically influences the distribution of these two elements in the wider area. The contents of these elements are very frequently concentrated as a result of the mineral raw materials processing. Kriging mapping indicated improving the anthropogenic inputs of significantly higher contents of these elements in separate zones, and also indicated the vertical/lithological distribution of these elements. TS/SS visualization strictly indicates areas with lithogenic enrichments of these elements. This is a specific association of elements with an isolated distribution in the areas of Neogene clastites, pyroclastites and volcanites. In this way the elements of this group are the main geogenic markers of old volcanism, represented in the Kratovo-Zletovo district (block).

Areal distribution of Factor 2 (Ni-Cr-Cu) is shown in Figure 6 (below). This geochemical association of elements shows a non-specific distribution. The distribution of this geogenic association of elements is specifically related to the geology of the area with dominant correlation with the Paleogene flysch and Paleozoic schists that are part of the dominant geological formations in the area. In this way the Ni-Cr-Cu association represents a typical geogenic marker of the area with noticeably high manifestation, but with high stability of the element content related to the above mentioned geologic formations.

Factor 3 (Ca-Sr-Al) dominates in areas with old (Neogene) and the young (Paleogene) volcanism (Figure 7). This is a specific association of elements with an isolated distribution in the areas of Neogene clastites, pyroclastites and volcanites. Dominance of this factor according to zones follows the following sequence: Z3 > Z4 > Z2 - Z1 (Figure 8).

Factor 4 (Fe-V-Mn-Mg) dominates in Zone 2, with strong lithogenic deposition in that area (Figure 7). Zone 2 presents an area with a dominant occurrence of Neogene pyroclastites and volcanites (with extension in Zone 1) and Rifeous shales.

The dominance of factors according to zones follows the following sequence Z1 > Z3 > Z2 - Z4 (Figure 8). In Figure 8 it can been seen that there are no significant differences of the factor patterns for topsoil and subsoil.

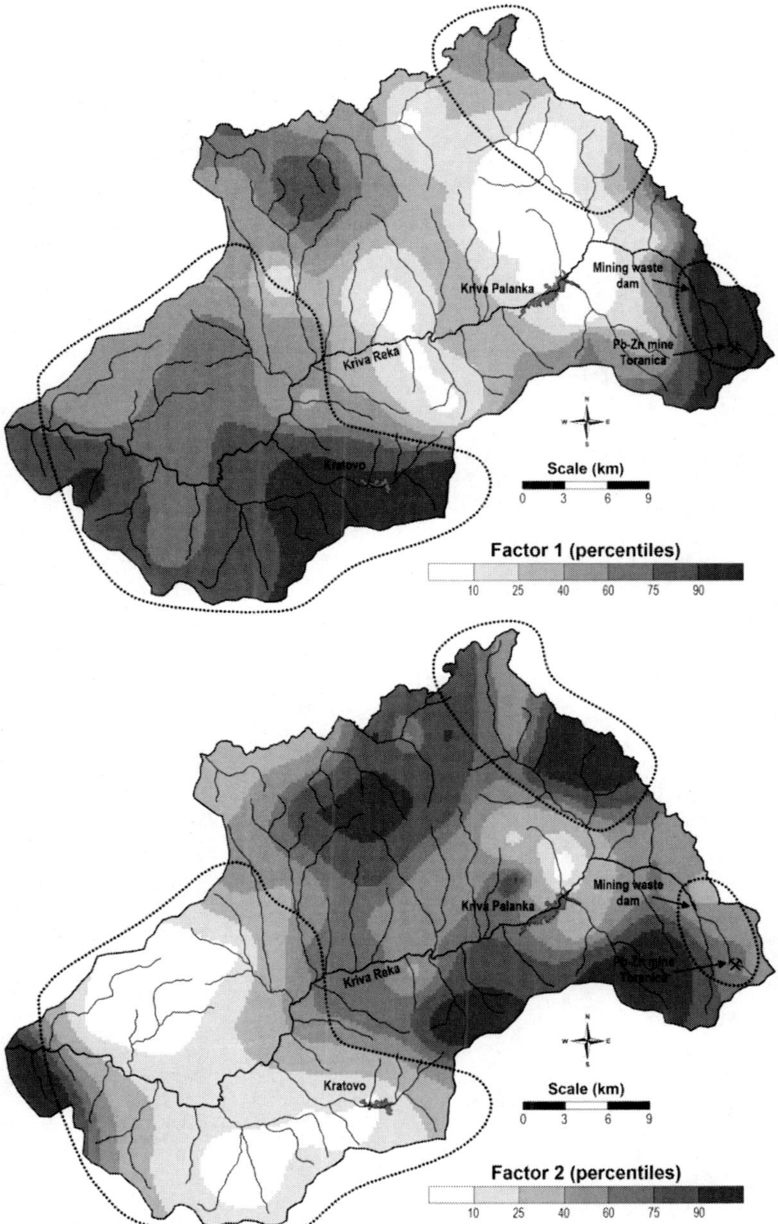

Figure 6. Spatial distribution of Factor 1 (above) and Factor 2 scores (below).

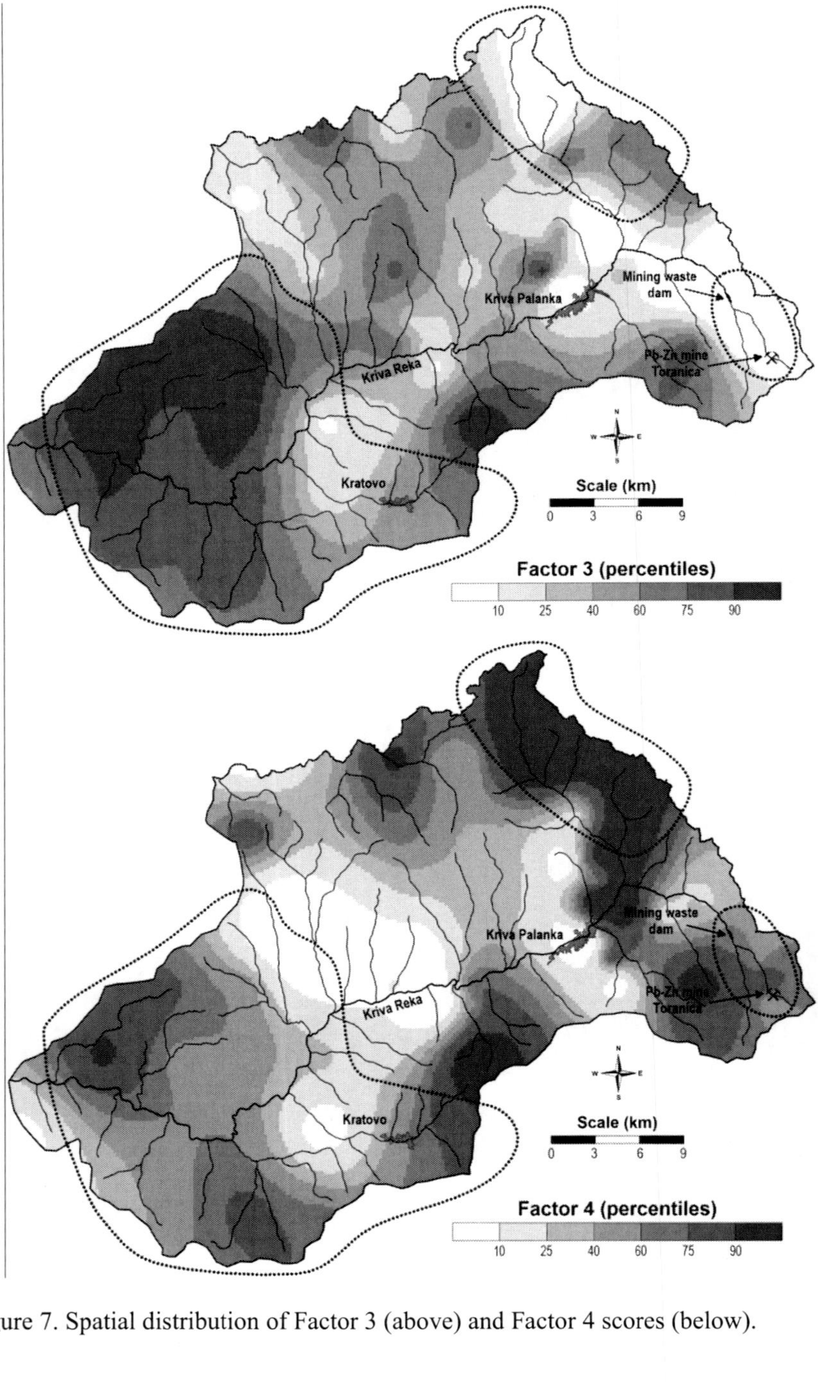

Figure 7. Spatial distribution of Factor 3 (above) and Factor 4 scores (below).

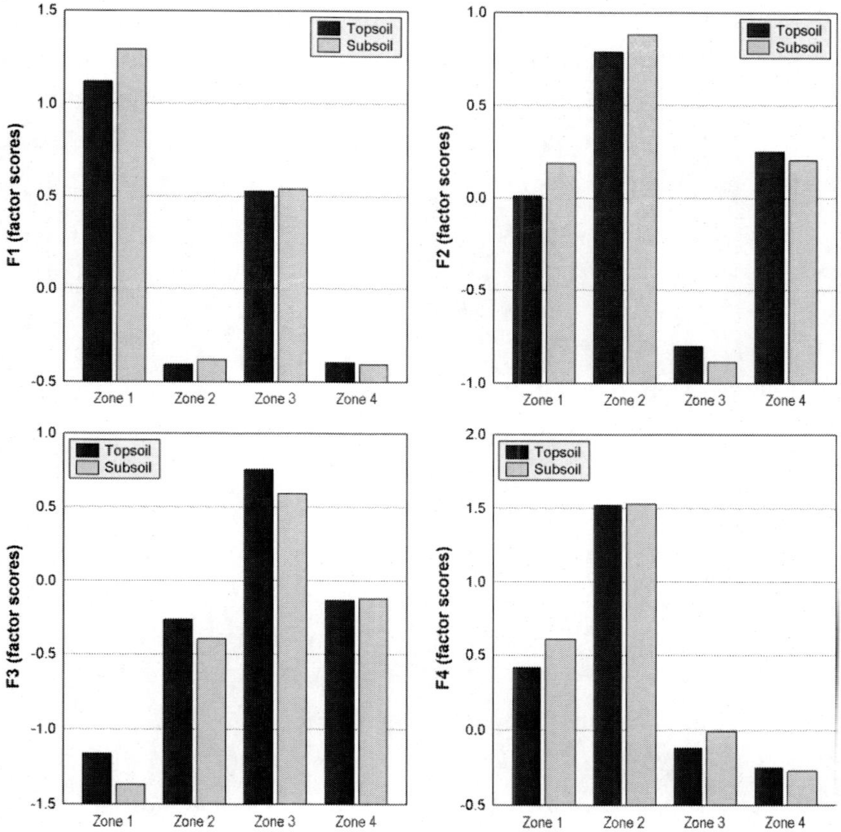

Figure 8. Distribution of factor scores (F1, F2, F3 and F4) in topsoil and subsoil regarding to determined zones.

CONCLUSION

Soil samples were used to monitor for the distribution of various chemical elements in the soil in the vertical direction at two levels (0-5 cm and 20-30 cm). No significant differences in the element content in the vertical direction were found, due to the characteristic lithogenic distribution. This confirms that external (anthropogenic) impact and environmental disasters do not disrupt the lithogenic distribution. The dominant geochemical associations, were F1 (Pb-Zn-As-Ba-K-Na), F2

(Ni-Cr-Cu), F3 (Ca-Sr-Al) and F4 (Fe-V-Mn-Mg). The natural distribution of 17 analyzed elements followed several lithogenic and anthropogenic anomalies specific to the region. The comparative analyses in the Kriva Reka River basin also suggest the effect of windblown dust on the element deposition, including local pollution due to Pb-Zn mining activities.

REFERENCES

Acton, Q. A., (2013). *Issues in environmental health and pollution.* Scholastic Edition Press, Atlanta, Georgia.

Agarwal, S. K., (2009). *Heavy metals pollution.* A P H Publishing Corporation, New Delhi.

Alloway, B. J., Ayres, D. C., (1997). *Chemical principles of environmental pollution*, (2nd ed.) Chapman & Hall, London.

Angelovska, S., Stafilov, T., Šajn, R., Balabanova, B., (2016). Geogenic and Anthropogenic Moss Responsiveness to Element Distribution Around a Pb–Zn Mine, Toranica, Republic of Macedonia. *Archives of Environmental Contamination and Toxicology,* 70(3), 487-505.

Artiola, J. F., Pepper, I., Brussean, L., (2004). *Environmental monitoring and characterization*, Elsevier Academic Press, San Diego.

Athar, M., Vohora, S., (1995). *Heavy metals and environment*, New Age International publishers, New Delhi.

Balabanova, B., Stafilov, T., Bačeva, K., Šajn, R., (2010). Biomonitoring of atmospheric pollution with heavy metals in the copper mine vicinity located near Radovıš, Republic of Macedonia. *Journal of Environmental Science and Health-Part A*, 45, 1504-1518.

Balabanova, B., Stafilov, T., Šajn, R., Bačeva, K., (2011). Distribution of chemical elements in attic dust as reflection of lithology and anthropogenic influence in the vicinity of copper mine and flotation. *Archives of Environmental Contamination and Toxicology*, 61, 173-184.

Balabanova, B., Stafilov, T., Šajn, R., Bačeva, K., (2012). Characterisation of heavy metals in lichen species *Hypogymnia physodes* and *Evernia prunastri* due to biomonitoring of air pollution in the vicinity of copper mine. *International Journal of Environmental Research*, 6, 779-794.

Balabanova, B., Stafilov, T., Šajn, R., Bačeva, K., (2013). Spatial distribution and characterization of some toxic metals and lithogenic elements in top soil and subsoil from copper mine environs. *International Journal of Environmental Protection,* 3, 1-9.

Balabanova, B., Stafilov, T., Šajn, R., Bačeva, K., (2014a) Variability assessment of metals distributions due to anthropogenic and geogenic impact in the lead-zinc mine and flotation "Zletovo" environs (moss biomonitoring). *Geologica Macedonica,* 28, 101-114.

Balabanova, B., Stafilov, T., Šajn, R., Bačeva, K., (2014b) Comparison of response of moss, lichens and attic dust to geology and atmospheric pollution from copper mine. *International Journal of Environmental Science and Technology*, 11 (2), 517-528.

Balabanova, B., Stafilov, T., Šajn, R., Tănăselia, C., (2016a). Multivariate factor assessment for lithogenic and anthropogenic distribution of macro and trace elements in surface water. Case study: basin of the Bregalnica river, Republic of Macedonia. *Macedonian Journal of Chemistry and Chemical Engineering*, 35(2), 1-16.

Balabanova, B., Stafilov, T., Šajn, R., Tănăselia, C., (2016b). Geochemical hunting of lithogenic and anthropogenic impacts on polymetallic distribution (Bregalnica river basin, Republic of Macedonia). *Journal of Environmental Science and Health, Part A.* 51(13) 1532-4117.

Balabanova, B., Stafilov, T., Šajn, R., Tănăselia, C., (2016c). Multivariate extraction of dominant geochemical markers for deposition of 69 elements in the Bregalnica River basin, Republic of Macedonia (moss biomonitoring). *Environmental Science and Pollution Research*, 23, 22852-22870.

Balabanova, B., Stafilov, T., Šajn, R., Tănăselia, C., (2017). Long-term geochemical evolution of lithogenic versus anthropogenic distribution of macro and trace elements in household attic dust. *Archives of Environmental Contamination and Toxicology*, 72(1), 88-107.

Barandovski, L., Stafilov, T., Šajn, R., Frontasyeva, M. V., Bačeva, K., (2013). Air pollution study in Macedonia using a moss biomonitoring technique, ICP-AES and AAS. *Macedonian Journal of Chemistry and Chemical Engineering*, 32(1), 89-107.

Boev, B., Yanev, Y., (2001). Tertiary magmatism within the Republic of Macedonia: A review. *Acta Vulcanologica*, 13, 57-71.

Box, G. E. P., Cox, D. R., (1964). An analysis of transformations. *Journal of the Royal Society, Series B (Methodological)*, 26(2), 211-252.

Dudka, S., Adriano, C. D., (1997) Environmental impacts of metal ore mining and processing: A review. *Journal of Environmental Quality,* 26, 590-602.

Hoenig, M., (2001). Preparation steps in environmental trace element analysis-facts and traps. *Talanta*, 54, 1021-1038.

ISO 14869-1:2001, Soil quality-Dissolution for the determination of total element content-Part 1: Dissolution with hydrofluoric and perchloric acids. *International Organization for Standardization*, Geneva, Switzerland.

Järup, L., (2003). Hazards of heavy metal contamination, *British Medical Bulletin*, 68, 167-182.

Kaymaz, I., (2005). Application of kriging method to structural reliability problems. *Structural Safety*, 27, 133-151.

Koljonen, T., (1992). *Geochemical Atlas of Finland, Part 2: Till.* Geological Survey of Finland, Espoo, Finland, 1992.

Krebs, R. E., (2006). *The history and use of our Earth's chemical elements: A reference guide*. Greenwood Press, Westport, CT.

Lazarevski, A., (1993). *Climate in Macedonia*, Kultura, Skopje, 1993.

Rakićević, T., (1968). Dumurdzanov N., Petkovski M. Basic geological map of SFRJ, sheet Štip, M 1:100,000 (map & interpreter), *Federal Geological Survey*, Beograd.

Reimann, C., Filzmoser, P., Fabian, K., Hron, K., Birke, M., Demetriades, A., Dinelli, E., Ladenberger, A., (2012). The concept of compositional data analysis in practice – Total major element concentrations in agricultural and grazing land soils of Europe. *Science of the Total Environment*, 426, 196-210.

Rühling, Å., Tyler, G., (1971). Regional differences in the heavy metal deposition over Scandinavia. *Journal of Applied Ecology*, 8, 497-507.

Rühling, Å., Tyler, G., (1973). Heavy metal deposition in Scandinavia. *Water, Air and Soil Pollution*, 2, 445-455.

Šajn, R., (2006). Factor analysis of soil and attic-dust to separate mining and metallurgy influence, Meza valley, Slovenia. *Mathematical Geology*, 38, 735-746.

Sakatam, S., Ashidam, F., Zakom, M., (2003). Structural optimization using Kriging approximation. *Computer Methods in Applied Mechanics and Engineering*, 192, 923-939.

Salminen, R., Batista, M. J., Bidovec, M., Demetriades, A., De Vivo, B., De Vos, W., Duris, M., Gilucis, A., Gregorauskiene, V., Halamic, J., Heitzmann, P., Jordan, G., Klaver, G., Klein, P., Lis, J., Locutura, J., Marsina, K., Mazreku, A., O'Connor, P. J., Olsson, S. Å., Rottesen, T., Petersell, V., Plant, J. A., Reeder, S., Salpeteur, I., Sandström, H., Siewers, U., Steenfelt, A., Tarvainen, T., (2005). *Geochemical Atlas of Europe,* Part 1, *Background Information, Methodology and Maps.* Geological Survey of Finland, Espoo.

Salomons, W., (1995). Environmental impact of metals derived from mining activities: Processes, predictions, preventions. *Journal of Geochemical Exploration*, 44, 5-23.

Sengupta, M., (1993). *Environmental impacts of mining: Monitoring, restoration and control*, Lewis Publishers, Boca Raton.

Serafimovski, T., Aleksandrov, M., (1995). Lead and zinc deposits and occurrences in the Republic of Macedonia. *Special edition of the Faculty of Mining and Geology*, Stip.

Serafimovski, T., Dolenec, T., Tasev, G., Dolenec, M., Rogan, N., (2007). Acid mine drainage systems and metal pollution around the active polymetallic mines in the eastern Macedonia. *Geologica Macedonica*, 21, 69-73.

Serafimovski, T., Tasev, G., Dolenec, T., (2006a) Petrological and geochemical features of the Neogene volcanites of the Osogovo Mountains, eastern Macedonia. *RMZ – Materials and Geoenvironment*, 52(3), 523-534.

Serafimovski, T., Tasev, G., Dolenec, T., (2006b). Actinolite-phengite-chlorite metasomatites from the Toranica Pb-Zn ore deposit in Macedonia. *RMZ - Materials and Geoenvironment*, 53(4), 445-453.

Siegel, F. R., (2002). *Environmental Geochemistry of potentially Toxic Metals*. Springer, Berlin, Heidelberg.

Stafilov, T., Bojkovska, R., Hirao, M., (2003). Air pollution monitoring system in the Republic of Macedonia. *Journal of Environment and Protection Ecology*, 4, 518–524.

Stafilov, T., Šajn, R., Pančevski, Z., Boev, B., Frontasyeva, M. V., Strelkova, L. P. (2008a). *Geochemical atlas of Veles and the environs*. Faculty of Natural Sciences and Mathematics Skopje.

Stafilov, T., Šajn, R., Boev, B., Cvetković, J., Mukaetov, D., Andreevski, M., (2008b). *Geochemical atlas of Kavadarci and the environs*. Faculty of Natural Sciences and Mathematics, Skopje.

Stafilov, T., Šajn, R., Boev, B., Cvetković, J., Mukaetov, D., Andreevski, M., Lepitkova, S., (2010b). Distribution of some elements in surface soil over the Kavadarci region, Republic of Macedonia. *Environmental Earth Sciences,* 61(7),1515-1530.

Stafilov, T., Šajn, R., Pančevski, Z., Boev, B., Frontasyeva, M. V., Strelkova, L. P., (2010a). Heavy metal contamination of surface soils around a lead and zinc smelter in the Republic of Macedonia. *Journal of Hazardous Materials,* 175, 896-914.

Stafilov, T., Balabanova, B., Šajn, R., Bačeva, K., Boev, B., (2010c). *Geochemical atlas of Radoviš and the environs and the distribution of heavy metals in the air*. Faculty of Natural Sciences and Mathematics, Skopje.

Stafilov, T., Šajn, R., Sulejmani, F., Bačeva, K., (2011). *Geochemical atlas of Kičevo and the environs.* Faculty of Natural Sciences and Mathematics, Skopje.

Stafilov, T., (2014). *Environmental pollution with heavy metals in the Republic of Macedonia.* Contributions, Section of Natural, Mathematical and Biotechnical Sciences, MASA, 35(2), 81–119.

Thornton, I., (1996). Impacts of mining on the environment; some local, regional and global issues. *Applied Geochemistry*, 11, 355-361.

Van het Bolcher, M., Van der Gon, D. H., Groenenberg, B. J., Ilyin, I., Reinds, G. J., Slootweg, J., Travnikov, O., Visschedijk, A., de Vries, W., (2006). *Heavy Metal Emissions, Depositions, Critical Loads and Exceedances in Europe*. Hettelingh, J. P., Sliggers, J. (eds.), National Institute for Public Health and the Environment, Bilthoven, the Netherlands.

VanLoon, G. W., Duffy, S. J., (2000). *Environmental Chemistry: A global perspective*, Oxford University Press, New York.

Wang, L. K., Chen, J. P., Hung, Y. T., Shammas, N. K., (2009). *Heavy metals in the Environment*. CRC Press Taylor & Francis Group, Boca Raton, FL.

Žibret, G., Šajn, R., (2010). Hunting for geochemical associations of elements: factor analysis and self-organizing maps. *Mathematical Geosciences*, 42, 681-703.

In: Soil Contamination
Editor: Jesper E. Lund
ISBN: 978-1-53613-266-3

Chapter 3

METAL OXIDE NANOPARTICLES: APPLICATIONS AND EFFECTS ON SOIL ECOSYSTEMS

***Vishnu D. Rajput*[1], *Tatiana Minkina*[1,*], *Alexey Fedorenko*[1,2], *Viktoriia Tsitsuashvili*[1], *Saglara Mandzhieva*[1], *Svetlana Sushkova*[1] and *Anatoly Azarov*[1]**

[1]Academy of Biology and Biotechnology,
Southern Federal University, Rostov, Russia
[2]Southern Scientific Center of Russian Academy of Sciences,
Rostov, Russia

ABSTRACT

Nanoparticles received great attention due to their unique properties and beneficiary applications. However, bioavailability and toxicity are

---

[*] Corresponding Author Email: tminkina@mail.ru.

key features for their massive employment. There are various types of nanoparticles but metal-based nanoparticles (e.g., silver, zinc oxide, titanium dioxide, iron oxide, copper oxide) are widely used in agriculture and allied sectors. It is predicted that soil ecosystems would be the final sink of various types of nanoparticles. So, it is very important to understand the behavior of nanoparticles in the soil and to evaluate the risk of nanoparticles in arable soil ecosystems or other real environmental scenarios. The soil microbiota and plants are among the major co-receptors of nanoparticles; especially, the soil-microbial biomass serves as a pool of nutrients and is a sensitive indicator of microbial changes in soils. Thus, it is more important to explore whether nanoparticles could compromise crop yield, soil properties, soil organisms and functional activities of the soil. Safe and sustainable application of nanotechnology in agriculture requires standards for the applications of nanoparticles in crop plants as well as the soil and water used for agricultural activity. The mechanism of nanoparticles toxicity remains unknown, however, it can be assumed to be closely related to the concentration, structure, size, surface area, and material used for nanoparticles synthesis. Studies conducted in hydroponic conditions do not reflect the interaction of nanoparticles with soil and soil microorganisms. However, real environmental studies should be conducted to observe toxicity of nanoparticles. The chapter will contribute to the better understanding of the fate and behaviour of nanoparticles in the soil, their uptake, and distribution within plants and impact of soil microbial communities as well as their interactions with other pollutants which is an essential prerequisite to environmentally realistic studies of nanoparticles ecotoxicology.

Keywords: copper, iron, microbial community, nanoparticles, plant, phytotoxicity, zinc

1. Introduction

The term "nanotechnology" was first used by Norio Taniguchi in 1974, describing dimensional accuracy (Taniguchi 1996). Nanotechnology is an interdisciplinary science associated with deliberately manufactured materials with at least one dimension less than 100 nm. Due to huge production and wide application of nanoparticles is affecting the whole ecosystem. At the same, nanoparticles are receiving great attention due to their unique properties and wide range of applications in agriculture industry, coatings, consumer products, cosmetics, chemical industries, electronics and optics, environmental remediation, food industries, fuel additives, textile industries, medical industries, paints, plastics, wastewater treatment and nano-fertilizers (Salata 2004; Minkina et al., 2010; Keller and Lazareva 2014; Servin et al., 2017; Weir et al., 2012; Yadav et al., 2014; Rajput et al., 2017a).

To bring out more types of nanoparticles with new characteristics and novel applications, research in this area is increasing tremendously and more & more money from public and private sectors is being invested. The Nano-era began in the early 2000s and global nanotechnology market in environmental applications reached \$23.4 billion in 2014 and \$41.8 billion by 2020 (BBC 2015), whereas, global nanocomposite market, in value term, should reach \$5.3 billion by 2021 from \$1.6 billion in 2016 (BBC 2017). An inventory of nanotechnology based consumer products introduced in the market lists 1,600+ products (Vance et al., 2015) and an online database (www.nanodb.dk) enlists 2,870 products in different categories containing nanoparticles or based on nanotechnology (Hansen et al., 2016). The actual figure of the global production of nanoparticles to the knowledge of authors is unavailable. An estimate in 2010 showed that 200 metric tons of nano-size Cu and CuO were produced (Zuverza-Mena et al., 2015). Zinc nanoparticles are estimated to be produced up to 5,500 tons per year replacing their bulk counterparts in a range of products (Connolly et al., 2016). Literature on the basis of data and modeling analysis indicates that the production, use and disposal of various metal based nanoparticles lead to release thousands of tons of most common nanoparticles (Ag, Al, Ce, Cu, Fe, Ti, Zn) into environment each year with majority of them end up in the soil, directly or through landfills from sludge and other waste (Minkina et al., 2010). Water and air is also getting the significantly high amount of nanoparticles share (Keller et al., 2013). Although estimated concentration of nanoparticles in the environment may not be entirely accurate as the values are indicated by predictive calculations (Sun et al., 2014).

Sources of nanoparticles entry into the environment are divided as natural and anthropogenic activities. Many anthropogenic objects and processes are sources of nanoparticles entry into the environment intentionally or accidentally. Nano copper is very toxic to aquatic life and, is used as fungicide and bactericide in agriculture (Zuverza-Mena et al., 2015). Zinc oxide nanoparticles are widely used in cosmetics industries (Nowack and Bucheli 2007).

Evidently, there are four major sources of nanoparticles emissions into the environment. Namely, manufacturing of raw materials, nano-enabled products, product use and waste management services (Tolaymat et al., 2017). Being released into the soil and water, nanoparticles can transport from one place to another and settle as soil sediments. Soil contaminated with nanoparticles could pollute groundwater (Tripathi et al., 2012). Once it end up into the soil ecosystems, it influences soil microbial diversity and plant performance (Kumar et al., 2012; Rajput et al., 2017a, b). Nanoparticles being highly reactive particles, can interact with other pollutants and form new compounds (Soni et al., 2015). The introduction of nanoparticles in the food chain via growing agricultural plants in contaminated soils could impose a serious threat to the human health (Rico et al., 2011).

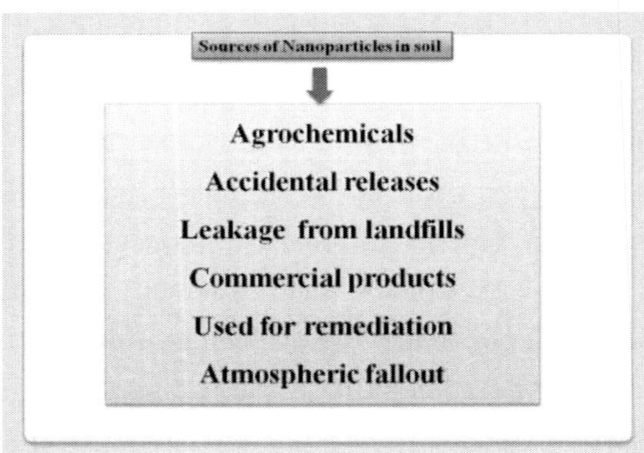

Figure 1. Release of nanoparticles in soil.

Phytotoxicity of nanoparticles is difficult to assess because it does not depend on only physical nature, production method, size and structure, but also on features of the biological models and quite possibly their growing technique (hydroponic, pot/model or in soil medium). Therefore, it is reasonable to speculate the consequences of the unregulated use of nanoparticles in agriculture and food industry.

2. Metallic Nanoparticles, Uptake, Translocation in the Plants and Phytotoxicity

The essential metallic oxide nanoparticles are chosen because they are essential elements for plants growth, and are nontoxic in a wide range of concentrations. This group of metals included Zn, Cu, Fe and Mn nanoparticles. Metal and metal oxide nanoparticles have potential of transformation or they get dissolved in soil because ZnO and CuO are very soluble. It is important to consider dissolution and transformation pathways for evaluating the efficacy or toxicity of nanoparticles. As nanoparticles are very small in size and have potential to enter, translocate, and penetrate physiological barriers to travel within the plant tissues, long-term effects of nanoparticles in plant systems are still unknown; however, some researchers suggests that the nanoparticles may cause morphological, physiological, genetic, and epigenetic changes which may alter plant growth and nutritional status (Table 1, 2, 3).

It is well understood that nanoparticles enter plant tissues either via root (root tips, rhizodermis, lateral root junctions) or the aboveground organs and tissues (cuticles, trichomes, stomata, stigma, and hydathodes) as well as through the wounds and root junctions. It has also been noticed that plants have different behavior towards excess metal ions present in the growth medium. Metal-tolerant plants may limit nanoparticles uptake into the photosynthetic tissue by restricting the transport across the root endodermis and storing them on the root cortex, whereas, hyper-accumulating plants accumulate excess amounts of nanoparticles in the harvestable tissues (Manceau et al., 2008).

Plant defence mechanism against the nanoparticles toxicity is not well explained. Generally, the phytotoxicity of nanoparticles is expressed as a chemical toxicity based on chemical composition, and stress stimuli caused by the surface, size, and shape of the nanoparticles. Phytotoxicity of any nanoparticle is largely influenced by its shape, size, chemical composition, and coating materials. Nanoparticles enter in root cells via symplastic or apoplastic pathways. The symplast pathway is where nanoparticle ion moves between cytoplasm/vacuoles of adjacent cells. However, the apoplast pathway absorbs nanoparticle ions in a certain way; near the xylem, the Casparian strip and translocate via cytoplasm. Through the symplastic and apoplastic pathways, nanoparticles translocate to the photosynthesis system. Zinc oxide nanoparticles internalization happens in the plant cells via apoplastic pathway. The schematic presentation of nanoparticles uptake via root and shoot is presented in figure 2.

It is also recommended that magnetic carbon-coated nanoparticles can easily penetrate root systems of cultivated crop plants by forming a biocompatible magnetic fluid known as bioferrofluid (Cifuentes et al., 2010). Essential metallic nanoparticles cause toxicity via dissolution and release of high concentration of essential ions, for example, Zn^{2+} and Cu^{2+} or the production of excess reactive oxygen species (ROS) (Letelier et al., 2010; Anjum et al., 2015; Minkina et al., 2016). ROS affects mitochondrial respiration and apoptosis, lipid peroxidation in the cell membrane, and induce oxidative stress in many plant species (Dimkpa et al., 2012; Shaw and Hossain 2013; Zhang et al., 2012).

Toxic effects of nanoparticles on plants are mainly expressed as suppressing the germination percentage, morphometric parameters such as root & shoot length, root hairs, biomass, bio-molecules, and cellular damages such as lipid peroxidation, protein damage, membrane destruction, chlorophyll content, transpiration and photosynthesis (Ghosh et al., 2015). Nanoparticles not only affect plant growth but also degrade crop quality, yield, nutrient level, starch content as well as soil properties and soil microbial community (Ben-Moshe et al., 2013; Le Van et al., 2016; Rajput et al., 2017a). The toxicity of nanoparticles might be related to the size of seeds, type of plant species, growth medium, stages and synthesis of nanoparticles (Lee et al., 2008).

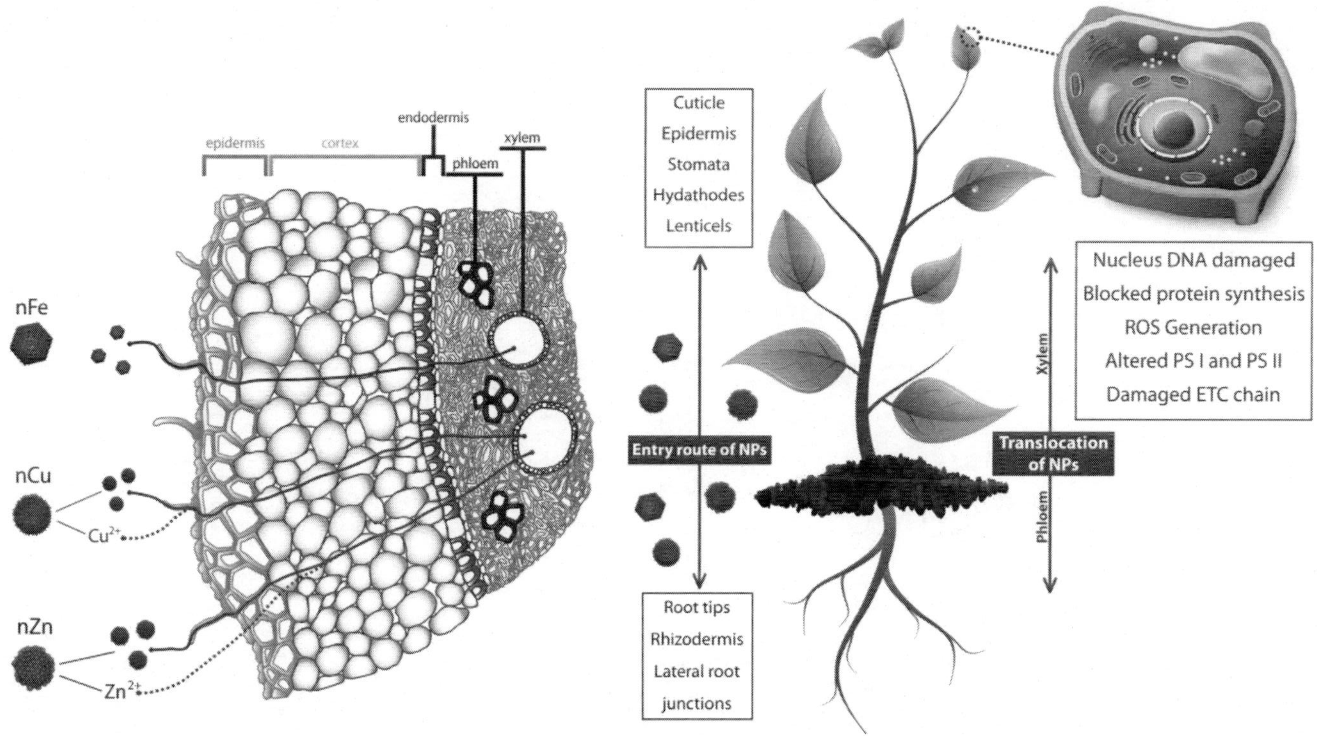

Figure 2. Schematic presentations of nanoparticles uptake, translocation and phytotoxicity.

Table 1. Effects of iron nanoparticles on plants growth

Sources	Plant species	Concentrations, mg L^{-1}	Toxic effects
Martinez-Fernandez et al., 2016	Sunflower	50-100	Reduced uptake and translocation of nutrients and root hydraulic conductivity
Martinez-Fernandez and Komarek 2016	*Tomato*	50-100	Iinhibited root hydraulic conductivity
Liu et al., 2016	*Lettuce*	1-50	Decreased root elongation
Gui et al., 2015	*Rice*	2-200	Iinhibited phyto-hormones
Trujillo-Reyes et al., 2014	Lettuce	10-20	Increased antioxidant enzyme activities, reduced root size, reduced chlorophylls content, affected hydraulic conductance
Wang et al., 2011	Ryegrass, Pumpkin	30-500	Increased root elongation, blocked aquaporins, oxidative stress
Sheykhbaglou et al., 2010	Soybean	0-1	Increased leaf and pod dry weight, 48% increase in grain yield

2.1. Iron Oxide Nanoparticles

Iron content is high in soil but a large proportion is fixed on soil particles, which is unavailable for plants due to its insoluble form (Fe^{3+}), as plant absorb Fe^{2+} form from the soil. Iron plays an important role in many physiological processes including chlorophyll biosynthesis, respiration, and redox reactions (Ye et al., 2015), and an essential micronutrient for plant growth. Iron is unavailable for plants in some soils, for instance, calcareous soil. Therefore, the application of iron nanoparticles has been increased recently. However, its long-term impacts on plants and soil ecosystems have not been studied yet. An increasing application of zero valent iron nanoparticles raises serious concerns for the environmental pollution. The zero-valent Fe nanoparticles have effectively been used in the treatment of a variety of environmental contaminants which increases the chances to its accumulation in the soil ecosystem. The studies show iron nanoparticles reduced the plant chlorophyll content, root hydraulic conductivity and affected transport of dissolved nutrients, especially Mg (Trujillo-Reyes et al., 2014; Martinez-Fernandez et al., 2016). The inhibitory effects of iron nanoparticles on the germination and growth (e.g., root, stem, leaf, and seedlings) on various crop plants are listed in Table 1.

Table 2. Effects of ZnO nanoparticles on plant performances

Sources	Plant species	Concentrations, mg L^{-1}	Toxic effects
Zafar et al., 2017	Mustard	500-1500	Affected seed germination and seedling growth
Priester et al., 2017	Soybean	0.05-0.5	Leaf chlorosis, necrosis and affected photo system II quantum efficiency
Singh and Kumar 2016	Spinach	1000	Reduced root and shoot length, total weight, chlorophyll and carotenoid content
Bandyopadhyay et al., 2015	Alfalfa	500-750	Reduced root and shoot biomass by 80%
Dimkpa et al., 2015	Bean	100-500	Inhibited growth, imbalanced nutrient in shoots, Na increased, Fe, Mn, Zn and Ca decreased
Yang et al., 2015	Maize, Rice	2000	Inhibited root elongation,
Chen et al., 2015	Rice		Induced phytotoxicity reduced ROS production and lipid peroxidation.
Mukherjee et al., 2014	Pea	25-500	Reduced chlorophyll and CAT content in leaves and APOX in roots and leaves
Yoon et al., 2014	Soybean	500	Reduced root & shoot growth, affected seed formation
Atha et al., 2012	Radish, Grasses	10-1000	Inhibited growth, damaged genetic material
Dimkpa et al., 2012	Wheat	500	Reduced root growth, increased lipid peroxidation and oxidized glutathione in roots, decreased chlorophyll content in shoots, increased ROS production
Zhao et al., 2012	Maize	20	Aggregates penetrated the root epidermis, cortex and accumulated in xylem vessels
Stampoulis 2009	Zucchini	1000	Reduced biomass (78–90%)
Lin and Xing 2007	Radish, Rapeseed, Ryegrass, Lettuce, Corn, Cucumber	2000	Reduced root growth and elongation

2.2. Zinc Oxide Nanoparticles

Zinc is an essential micronutrient but excess amount may cause chlorosis and inhibit plant growth (Vallee and Auld 1990; Schutzendubel and Polle 2002). Zinc nanoparticles are spherical or polished metal particles with a high specific surface area with molecular formula ZnO. Zinc nanoparticles are widely used in ceramics, glass, cement, rubber, lubricants, paints, ointments, adhesives, plastics, sealants, pigments, foods and battery industries. Several methods are used to synthesise ZnO nanostructures, nano-particles, nano-wires, nano-rods, nano-tubes, nano-belts, and other complex morphologies. The negative impacts of ZnO nanoparticles on plant growth and development have been reported in several plants such as soybean, wheat, cucumber, mustard, alfalfa, radish, ryegrass, lettuce, maize, zucchini, spinach etc (Rajput et al., 2017a, b, c) (Table 2).

2.3. Copper Oxide Nanoparticles

Copper is widely distributed in plant tissues, and is an essential micronutrient for the plant growth, involved in many physiological processes, and widely used in agriculture industries, cosmetics, coatings, environmental remediation, food industries, chemical industries, fuel additives, textile industries, medical industries, paints, plastics, wastewater treatment and electronics (Rafique et al., 2017). It is reported that on average 1 kg of dry plant tissue contain approximately 10 mg of Cu (Baker and Senef 1995). On the one side, it has reshaped current science by being utilized in different industries, but on the other hand exerted toxic effects on the living systems (Minkina et al., 2015; Burachevskaya et al., 2017). It has also been reported that low concentration of CuO nanoparticles is more soluble and more toxic to aquatic organisms (Aruoja et al., 2009). Copper has an outstanding electrical conductivity, good catalytic behaviour and surface-enhanced Raman scattering activity; hence, it attracted the attention of researchers for using it as an essential component in the future nano-devices.

Table 3. Effects of CuO nanoparticles on plants performances

Sources	Plant species	Concentrations, $mg\ L^{-1}$	Toxic effects
Olchowik et al., 2017	Oak	0-50	Disturbed shape, plastoglobules and starch contents of leaves cells
Rajput et al., 2017c	Spring barley	10000	Affected germination rate, retarded root and shoot length
Singh and Kumar 2016	Spinach	1000	Reduced root length, shoot length, total weight, chlorophyll, and carotenoid content
Nhan et al., 2016	Cotton	1000	Height and root length decreased after 10 day exposure in *in-vitro*
Ebbs et al., 2016	Carrot	1-1000	Biomass decreased significantly
Le Van et al., 2016	Cotton	10-1000	Reduced the uptake of minerals (B, Mo, Mn, Mg, Zn, Fe)
Deng et al., 2016	Onion	0-80	Reduced/stopped growth of root length and showed deformation on root cap surface and meristematic zone
Rao and Shekhawat 2016	Mustard	0-500	Reduced shoot growth, root shortening
Hong et al., 2015	Lettuce, Alfalfa	0-1000	Reduced root growth, decreased P, S, Cu, and Fe in shoots, increased APX in roots
Dimkpa et al., 2015	Bean	100-500	Inhibited growth, imbalanced metal nutrient in shoots, increased Na, decreased Fe, Mn, Zn and Ca
Yang et al., 2015	Maize, Rice	2000 40-80	Inhibited root elongation, reduced shoot length
Zuverza-Mena et al., 2015	Cilantro	0-100	Affected germination, reduced shoot elongation
Nair and Chung 2015	Mustard	0-1500	Root observed thinner and more brittle with brown necrotic lesions. Affected antioxidant activities
Nair and Chung 2014a	Soybean	50-500	Reduced shoot and root growth, weight, and total chlorophyll content
Nair et al., 2014b	Mungbean	200-500	Reduced root length, biomass and affected primary and lateral root growth, and total chlorophyll contents
Shaw et al., 2014	Barley	0-1.5	Induced production of ROS, damaged membrane, affected shoot and root growth
Moon et al., 2014	Cucumber	100-600	Inhibited seed germination and root elongation
Wang et al., 2012	Maize	2-100	Chlorotic symptoms, reduced biomass and root elongation
Dimkpa et al., 2012	Wheat	500	Shoot length reduced significantly by 13% and root length reduced 59%, brown necrotic lesions on roots, decreased chlorophyll levels
Dimkpa et al., 2012	Wheat	500	Inhibited root and shoot growth, produced oxidative stress
Atha et al., 2012	Radish, Grasses	10-1000	Affected growth and damaged nucleic acid
Lee et al., 2008	Mungbean, Wheat	200-1000	Reduced plant biomass

A bacterium belonging to the genus *Serratia* isolated from an insect *Stibara* species was first micro-organism reported for synthesis of Cu nanoparticles of 10-30 nm using 5 mM copper sulphate, and 15 and 40 nm synthesized by using *Calotropis procera* latex and flowers broth of *Aloe vera* (Harne et al., 2012; Karimi and Mohsenzadeh 2015). Addition of *Centella asiatica* leaves extract to copper acetate formed copper nanoparticles ranging from 2 to 5 nm (Devi and Singh 2014).

Recent studies of CuO nanoparticles toxicity show negative impacts on seed germination and plant growth on various crop plants i.e., lettuce, alfalfa, wheat, mung bean, kidney-bean, maize, cilantro, rice, spinach, onion, mustard, barley, cotton, chickpea, radish, pistia and oak (Table 3) (Rajput et al., 2017b).

3. Effects of Nanoparticles on Microbial Community

Nanoparticles can enter soil through various pathways, such as agricultural amendments of sewage sludge, atmospheric deposition, landfills, or accidental spills during industrial production (Simonin and Richaume 2015). At present inadequate information is available on how nanoparticles affect the soil microbial community. They may affect directly by changes in the bioavailability of other toxins and nutrients or indirectly via interactions with natural organic compounds possible interactions with toxic organic compounds which may increase or decrease toxicity of nanoparticles (Haris and Iqbal 2017). Nanoparticles and their ions can produce free radicals which can damage microbial cell membrane, DNA, and mitochondria resulting in death of microbes (Rajput et al., 2017a, b). The impact of nanoparticles on microbial activity could be determined through measurement of soil respiration and enzymatic activities (Simonin and Richaume 2015). Metallic nanoparticles could influence soil enzymatic activities and changes in soil bacterial abundance (Figure 3).

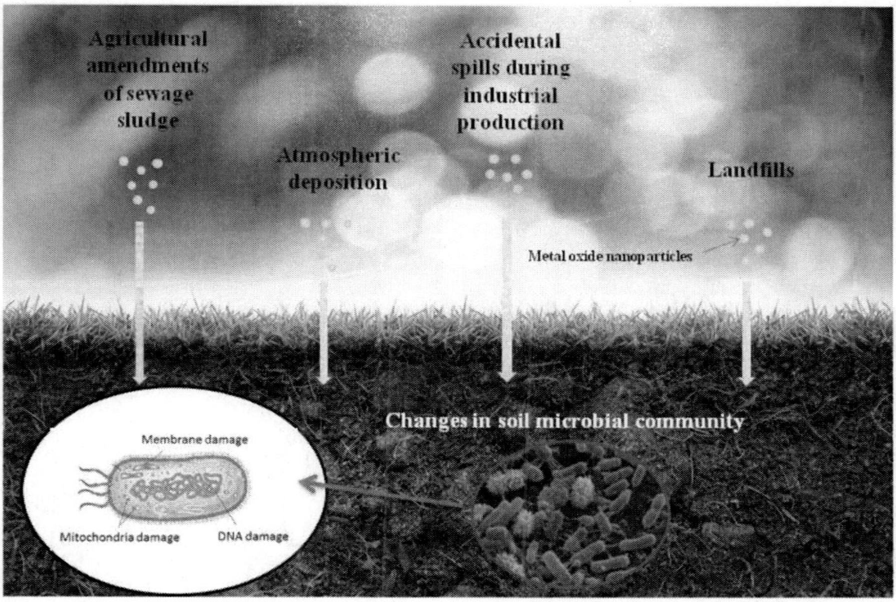

Figure 3. Effects of nanoparticles on soil microbial community.

You et al., (2017) suggested that the soil types could play an important role in nanoparticles toxicity. Moreover, the soil type is a key component for nanoparticles toxicity on soil bacterial community composition and size. Nanoparticles ions are considered as a major cause of lethality to both the pathogenic and beneficial microbes (Lofts et al., 2013) and very toxic to the native soil bacteria (Concha-Guerrero et al., 2014). The study on the toxic effect of nanoparticles is not only limited for beneficial soil microbe such as nitrifying bacteria, nitrogen-fixing bacteria, *arbuscular mycorrhiza* and *rhizobacteria*, however it also influences other soil microbes. Recent studies showed that nanoparticles may affect enzymatic activities, metabolic activities, nitrification potential, colony count and abundances of soil bacteria diversity (Ge et al., 2012; Peyrot et al., 2014; Colman et al., 2013; He et al., 2016; Juan et al., 2017; Rajput et al., 2017a; McGee et al., 2017). The numbers of studies showing the effects of nanoparticles on the microbial community are included in Table 4.

Table 4. Effects of various nanoparticles on soil microbial communities

Sources	Nanoparticles	Concentrations in soil, mg kg^{-1}	Toxic effects
You et al., 2017	ZnO, Fe_3O_4	500-1000, 2000	Affected enzymatic activities (invertase, urease, catalase, and phosphatase) and bacterial communities of saline-alkali and black soils
Jiling et al., 2016	Fe_3O_4	0.1-10.0	Significantly decreased the content of bacteria in soil
Chai et al., 2015	ZnO	1000	Affected plate counts of *Azotobacter*, *P-solubilizing* and *K-solubilizing* and inhibited enzymatic activities
Xu et al., 2015	CuO	0-1000	Decreased soil microbial biomass, enzymatic activities, disturbed community structures
Shah et al., 2014	Fe	550	Individual analysis showed effect on bacterial group
Ben-Moshe et al., 2013	CuO	10	Affected soil microbial community
Kumar et al., 2012	Cu	220	Reduced C and N biomass, disturbed microbial community structure
Ge et al., 2011	ZnO	500-2000	Altered soil bacterial community structure and decreased bacterial diversity

CONCLUSION AND FUTURE PROSPECTIVES

It is an essential prerequisite to environmentally realistic studies of the ecotoxicology of metallic nanoparticles. A series of safety evaluation and toxicological risk assessment standards must be formulated, including exposure route, safe dose, and synthesis methods before applying nanoparticles in the environment. Future studies should address questions such as the impact of nanoparticles on food quality, soil ecosystem and should define the threshold limits. Data should be generated from appropriately contaminated area because most of the reported studies have been performed in hydroponic conditions, model experiments, or artificially contaminated soil. It is anticipated that the chapter could stimulate the further researches on nanoparticles toxicity especially on soil ecosystems.

Acknowledgments

This work is supported by the Russian Science Foundation (no. 16-14-10217).

References

Anjum, N. A., Adam, V., Kizek, R., Duarte, A. C., Pereira, E., Iqbal, M., Lukatkin, A. S., Ahmad, I. (2015). Nanoscale copper in the soil-plant system toxicity and underlying potential mechanisms. *Environ. Res.* 138, 306-325.

Aruoja, V., Dubourguier, H. C., Kasemets, K., Kahru, A. (2009). Toxicity of nanoparticles of CuO, ZnO and TiO_2 to microalgae *Pseudokirchneriella subcapitata. Sci. Total Environ.* 407, 1461-1468.

Atha, D. H., Wang, H., Petersen, E. J., Cleveland, D., Holbrook, R. D., Jaruga, P., Dizdaroglu, M., Xing, B., Nelson, B. C. (2012). Copper oxide nanoparticle mediated DNA damage in terrestrial plant models. *Environ. Sci. Technol.* 46, 1819-1827.

Baker, D. E., Senef, J. P. (1995). Copper in: Alloy, B.J. (Ed.), *Heavy metals in soils.* London, Blackie academic and professional.

Bandyopadhyay, S., Plascencia-Villa, G., Mukherjee, A., Rico, C. M., José-Yacamán, M., Peralta-Videa, J. R., Gardea-Torresdey, J. L. (2015). Comparative Phytotoxicity of ZnO NPs, Bulk ZnO, and Ionic Zinc Onto the Alfalfa Plants Symbiotically Associated with *Sinorhizobium meliloti* in Soil. *Sci. Total Environ.* 515-516, 60-69.

BBC. (2015) *Nanotechnology in environmental applications: the global market*. Nano, 39C. https://www.bccresearch.com/market-research/nanotechnology.

BBC. (2017). *Global markets for nanocomposites, nanoparticles, nanoclays, and nanotubes*. Nano, 21G. https://www.bccresearch.com/market-research/nanotechnology.

Ben-Moshe, T., Frenk, S., Dror, I., Minz, D., Berkowitz, B. (2013). Effects of metal oxide nanoparticles on soil properties. *Chemosphere*. 90(2), 640-646.

Chai, H., Yao, J., Sun, J., Zhang, C., Liu, W., Zhu, M., Ceccanti, B. (2015). The effect of metal oxide nanoparticles on functional bacteria and metabolic profiles in agricultural soil. *Bull Environ Contam Toxicol.* 94, 490-495.

Chen, J., X. Liu, C. Wang, S. S., Yin, X. L., Li, W. J., Hu, M., Simon, et al., (2015). Nitric oxide ameliorates zinc oxide nanoparticles induced phytotoxicity in rice seedlings. *J. Hazard Mater.* 297, 173-182.

Cifuentes, Z., Custardoy, L., de la Fuente, J., Marquina, C., Ibarra, M., Rubiales, D., et al., (2010). Absorption and translocation to the aerial part of magnetic carbon-coated nanoparticles through the root of different crop plants. *J Nanobiotechnology*. 8(1), 26-33.

Colman, B. P., Arnaout, C. L., Anciaux, S., Gunsch, C. K., Hochella, M. F. Jr, Kim, B., Lowry, G. V., McGill, B. M., Reinsch, B. C., Richardson, C. J., Unrine, J. M., Wright, J. P., Yin, L., Bernhardt, E. S., (2013). Low concentrations of silver nanoparticles in sewage sludge cause adverse ecosystem responses under realistic field scenario. *PLoS One*. 8, 57189.

Concha-Guerrero, S. I., Brito, E. M. S., Piñón-Castillo, H. A., et al., (2014). Effect of CuO nanoparticles over isolated bacterial strains from agricultural soil. *J. Nanomater.* 2014, 13.

Connolly, M., Fernández, M., Conde, E., Torrent, F., Navas, J. M., Fernández-Cruz, M. L. (2016). Tissue distribution of zinc and subtle oxidative stress effects after dietary administration of ZnO nanoparticles to rainbow trout. *Sci Total Environ*. 551-552, 334- 343.

Deng, F., Wang, S., Xin, H. (2016). Toxicity of CuO nanoparticles to structure and metabolic activity of *Allium cepa* root tips. *Bull. Environ. Contam. Toxicol*. 97, 702-708.

Devi, H. S., Singh, T. D. (2014). Synthesis of copper oxide nanoparticles by a novel method and its application in the degradation of methyl orange. *Adv. Electron. Electr. Eng.* 4, 83- 88.

Dimkpa, C. O., McLean, J. E., Britt, D. W., Anderson, A. J. (2015). Nano-CuO and inter action with nano-ZnO or soil bacterium provide evidence for the interference of nanoparticles in metal nutrition of plants. *Ecotoxicology*. 24, 119-129.

Dimkpa, C. O., McLean, J. E., Latta, D. E., Manangón, E., Britt, D. W., Johnson, W. P., Boyanov, M. I., Anderson, A. J. (2012). CuO and ZnO nanoparticles: phytotoxicity, metal speciation, and induction of oxidative stress in sand-grown wheat. *J. Nanopart. Res.* 14, 1-15.

Ebbs, S. D., Bradfield, S. J., Kumar, P., White, J. C., Musante, C., Ma, X. (2016). Accumulation of zinc, copper, or cerium in carrot *(Daucus carota)* exposed to metal oxide nanoparticles and metal ions. *Environ. Sci. Nano*. 3, 114-126.

Ge, Y. G., Schimel, J. P., Holden, P. A. (2011). Evidence for negative effects of TiO_2 and ZnO nanoparticles on soil bacterial communities. *Environ. Sci. Technol.* 45, 1659-1664.

Ge, Y. G., Schimel, J. P., Holden, P. A. (2012). Identification of soil bacteria susceptible to TiO_2 and ZnO nanoparticles. *Appl. Environ. Microbiol*. 78, 6749-6758.

Ghosh, M., Bhadra, S., Adegoke, A., Bandyopadhyay, M., Mukherjee, A. (2015). MWCNT uptake in Allium cepa root cells induces cytotoxic and genotoxic responses and results in DNA hypermethylation. *Mutat. Res. Fundam. Mol. Mech. Mutagen*. 774, 49-58.

Gui, X., Deng, Y. Q., Rui, Y. K., Gao, B. B., Luo, W. H., Chen, S. L., Nhan, L. V., Li, X. G., Liu, S. T., Han, Y. N., et al., (2015). Response difference of transgenic and conventional rice (*Oryza sativa*) to nanoparticles (gamma Fe_2O_3). *Environ. Sci. Pollut. Res*. 22, 17716- 23.

Hansen, S. F., Heggelund, R. L., Besora, P. R., Mackevica, A., Boldrin, A., Baun, A. (2016). Nanoproducts-what is actually available to European consumers? *Environ. Sci. Nano.* 3, 169-180.

Haris, Z. and Iqbal, A. (2017). Impact of metal oxide nanoparticles on beneficial soil microorganisms and their secondary metabolites. *Int. J. Life Sci. Scienti. Res.* 3(3), 1020- 1030.

Harne, S., Sharma, A., Dhaygude, M., Joglekar, S., Kodam, K., Hudlikar, M. (2012). Novel route for rapid biosynthesis of copper nanoparticles using aqueous extract of *Calotropis procera L.* latex and their cytotoxicity on tumor cells. *Colloids Surf. B Biointerfaces*. 95, 284-288.

He, S., Feng, Y., Ni, J., Sun, Y., Xue, L., Feng, Y., Yu, Y., Lin, X., Yang, L. (2016). Different responses of soil microbial metabolic activity to silver and iron oxide nanoparticles. *Chemosphere*. 147, 195-202.

Hong, J., Rico, C. M., Zhao, L., Adeleye, A. S., Keller, A. A., Peralta-Videa, J. R., GardeaTorresdey, J. L. (2015). Toxic effects of copper-based nanoparticles or compounds to lettuce *(Lactuca sativa)* and alfalfa *(Medicago sativa). Environ. Sci. Process Impact.* 17, 177-185.

Jiling, C., Youzhi, F., Xiangui, L., Junhua, W. (2016). Arbuscular mycorrhizal fungi alleviate the negative effects of iron oxide nanoparticles on bacterial community in rhizospheric soils. *Front. Environ. Sci.* 4, 10.

Juan, W., Kunhui, S., Li, Z. and Youbin, S. (2017). Effects of Silver Nanoparticles on Soil Microbial Communities and Bacterial Nitrification in Suburban Vegetable Soils. *Pedosphere*. 27(3), 482-490.

Karimi, J. and Mohsenzadeh, S. (2015). Rapid, green, and eco-friendly biosynthesis of copper nanoparticles using flower extract of *Aloe vera. Synth. React. Inorg. Met. Org. Nano Met. Chem*. 45, 895-898.

Keller, A. A. and Lazareva, A. (2014). Predicted releases of engineered nanomaterials: from global to regional to local. *Environ Sci Technol Lett*. 1, 65-70.

Keller, A. A., McFerran, S., Lazareva, A., Suh, S. (2013). Global life cycle releases of engineered nanomaterials. *J. Nanopart. Res.* 15, 1692.

Kumar, N., Shah, V., Walker, V. K. (2012). Influence of a nanoparticle mixture on an arctic soil community. *Environ. Toxicol. Chem.* 31, 131-135.

Le Van, N., Ma, C. X., Shang, J. Y., Rui, Y. K., Liu, S. T., Xing, B. S. (2016). Effects of CuO nanoparticles on insecticidal activity and phytotoxicity in conventional and transgenic cotton. *Chemosphere.* 144, 661-670.

Lee, W. M., An, Y. J., Yoon, H., Kweon, H. S. (2008). Toxicity and bioavailability of copper nanoparticles to the terrestrial plants mungbean *(Phaseolus radiatus)* and wheat *(Triticum aestivum)*: Plant agar test for water-insoluble nanoparticles. *Environ. Toxicol. Chem.* 27, 1915-1921.

Letelier, M. E., Sanchez-Jofre, S., Peredo-Silva, L., Cortes-Troncoso, J., Aracena Parks, P., (2010). Mechanisms underlying iron and copper ions toxicity in biological systems: pro- oxidant activity and protein-binding effects. *Chem-Biol. Interact.* 188, 220-227.

Lin, D. and Xing, B. (2007). Phytotoxicity of nanoparticles: inhibition of seed germination and root growth. *Environ. Pollut.* 150, 243-250.

Liu, R. Q., Zhang, H. Y., Lal, R. (2016). Effects of stabilized nanoparticles of copper, zinc, manganese, and iron oxides in low concentrations on lettuce (*Lactuca sativa*) seed germination: nanotoxicants or nanonutrients? *Water, Air, & Soil Pollution.* 227, 1-14.

Lofts, S., Criel, P., Janssen, C. R., Lock, K., McGrath, S. P., Oorts, K., Rooney, C. P., Smolders, E., Spurgeon, D. J., Svendsen, C., Eeckhout, H. V., Zhao, F. Z. (2013). Modelling the effects of copper on soil organisms and processes using the free ion approach: Towards a multi-species toxicity model. *Environ. Pollut.* 178, 244-253.

Manceau, A., Nagy, K. L., Marcus, M. A., Lanson, M., Geoffroy, N., Jacquet, T., Kirpichtchikova, T. (2008). Formation of metallic copper nanoparticles at the soil-root interface. *Environ. Sci. Technol.* 42, 1766-1772.

Martinez-Fernandez, D., Barroso, D., Komarek, M. (2016). Root water transport of *Helianthus annuus L.* under iron oxide nanoparticle exposure. *Environ. Sci. Pollut. Res.* 23, 1732-41.

Martinez-Fernandez, D., Komarek, M. (2016). Comparative effects of nanoscale zero-valent iron (nZVI) and Fe_2O_3 nanoparticles on root hydraulic conductivity of *Solanum lycopersicum L. Environ. Exp. Bot.* 131, 128-36.

McGee, C. F., Storey, S., Clipson, N., Doyle, E. (2017). Soil microbial community responses to contamination with silver, aluminium oxide and silicon dioxide nanoparticles. *Ecotoxicology*. 1-10.

Minkina, T. M., Motusova, G. V., Nazarenko, O. G., Mandzhieva, S. S. (2010). *Heavy metal compounds in soil: transformation upon soil pollution and ecological significance*. N.Y.: Nova Science Publishers, Inc., 189 p. ISBN: 978-1-60876-466-2.

Minkina, T. M., Bauer, T. V., Batukaev, A. A., Mandzhieva, S. S., Burachevskaya, M. V., Sushkova, S. N., Varduni, T. V., Sherstnev, A. K., Kalinichenko, V. P. (2015). Transformation of technogenic Cu and Zn compounds in chernozem. *Environ. Eng. Manag. J.* 14, 481-486.

Minkina, T. M., Nevidomskaya, D. G., Shuvaeva, V. A., Soldatov, A. V., Tsitsuashvili, V. S., Zubavichus, Y. V., Rajput, V. D., Burachevskaya, M. V. (2016) Studying the transformation of Cu^{2+} ions in soils and mineral phases by the XRD, XAFS, and sequential fractionation methods. *J. Geochem. Explor*. http://dx.doi.org/10.1016/j.gexplo.2016.10.007.

Moon, Y. S., Park, E. S., Kim, T. O., Lee, H. S., Lee, S. E. (2014). SELDI-TOF MS-based discovery of a biomarker in Cucumis sativus seeds exposed to CuO nanoparticles. *Environ. Toxicol. Phar.* 38, 922-931.

Mukherjee, A., Peralta-Videa, J. R., Bandyopadhyay, S., Rico, C. M., Zhao, L., and Gardea- Torresdey, J. L. (2014). Physiological Effects of Nanoparticulate ZnO in Green Peas (*Pisum sativum* L.) Cultivated in Soil. *Metallomics.* 6 (1), 132-138.

Nair, P. M. G., Chung, I. (2014a). A mechanistic study on the toxic effect of copper oxide nanoparticles in soybean *(Glycine max L.)* root development and lignification of root cells. *Biol. Trace Element Res.* 162, 342-352.

Nair, P. M. G., Chung, I. M. (2015). Study on the correlation between copper oxide nanoparticles induced growth suppression and enhanced lignification in Indian mustard *(Brassica juncea L.). Ecotoxicol. Environ. Saf.* 113, 302-313.

Nair, P. M. G., Kim, S. H., Chung, M. (2014b). Copper oxide nanoparticle toxicity in mung bean *(Vigna radiata L.)* seedlings: Physiological and molecular level responses of in vitro grown plants. *Acta Physiol. Plant.* 36, 2947-2958.

Nhan, L. V., Yukui, R., Weidong, C., Jianying, S., Shutong, L., Trung, N. Q., Liming, L. (2016). Toxicity and bio-effects of CuO nanoparticles on transgenic Ipt-cotton. *J. Plant Interact.* 11, 108-116.

Nowack, B. and Bucheli, T. D. (2007). Occurrence, behavior and effects of nanoparticles in the environment. *Environ. Pollut.* 150(1), 5-22.

Olchowik, J., Bzdyk, R. M., Studnicki, M., Bederska-Błaszczyk, M., Urban, A., Aleksandrowicz-Trzcińska, M. (2017). The effect of silver and copper nanoparticles on the condition of english oak *(Quercus robur L.)* seedlings in a container nursery experiment. *Forests.* 8, 310.

Peyrot, C., Wilkinson, K. J., Desrosiers, M., Sauvé, S. (2014). Effects of silver nanoparticles on soil enzyme activities with and without added organic matter. *Environ. Toxicol. Chem.* 33, 115-125.

Priester, J. H., Shelly, C. M., Katherine, E., Yuan, G., Ying, W., Roger, M. N., Joshua, P. S., Goggi, S., Gardea-Torresdey, J. L., Patricia, A. H. (2017). Damage assessment for soybean cultivated in soil with either CeO_2 or ZnO manufactured nanomaterials. *Sci. Total Environ.* 579, 1756-1768.

Rafique, M., Shaikh, A. J., Rasheed, R., Tahir, M. B., Bakhat, H. F., Rafique, M. S., Rabbani, F. (2017). A review on synthesis, characterization and applications of copper nanoparticles using green method. *Nano*. 12, 04.

Rao, S., Shekhawat, G. S. (2016). Phytotoxicity and oxidative stress perspective of two selected nanoparticles in *Brassica juncea. 3 Biotech*. 6, 244.

Rajput, V. D., Minkina, T., Suskova, S., et al., (2017a). Effect of nanoparticles on crops and soil microbial communities. *J. Soils Sediments.* 1-9. doi 10.1007/s11368-017-1793-2.

Rajput, V. D., Minkina, T., Suskova, S., Mandzhieva, S., Tsitsuashvili, V., Chapligin, V., Fedorenko, A. (2017b). Effects of copper nanoparticles

(CuO NPs) on crop plants: a mini review. *Bio Nano Science*. DOI: 10.1007/s12668-017-0466-3.

Rajput, V. D., Tstitsuashvili, V. S., Sushkova, S. N., Nevidomskaya, D. G. (2017c). Effects of ZnO and CuO nanoparticles on soil, plant and microbial community. *International Scientific Conference XX Dokoutchaev Youth Readings, St. Petersburg, Russia*. 12.

Rico, C. M., Majumdar, S., Duarte-Gardea, M., Peralta-Videa, J. R., Gardea-Torresdey, J. L. (2011). Interaction of nanoparticles with edible plants and their possible implications in the food chain. *J. Agric. Food Chem*. 59, 3485-3498.

Salata, O. V. (2004). Applications of nanoparticles in biology and medicine. *J. Nano-biotech*. 2, 3.

Schutzendubel, A. and Polle, A. (2002). Plant responses to abiotic stresses: heavy metal-induced oxidative stress and protection by mycorrhization. *J. Exp. Bot*. 53, 1351-1365.

Servin, A. D., De la Torre-Roche, R., Castillo-Michel, H., Pagano, L., Hawthorne, J., Musante, C., Pignatello, J., Uchimiya, M., White, J. C. (2017). Exposure of agricultural crops to nanoparticle CeO_2 in biochar-amended soil. *Plant Physiol. Biochem*. 110, 147-157.

Shah, V., Collins, D., Walker, V. K., Shah, S. (2014). The impact of engineered cobalt, iron, nickel and silver nanoparticles on soil bacterial diversity under field conditions. *Environ. Res. Lett*. 9(2), 024001.

Shaw, A. K., Ghosh, S., Kalaji, H. M., Bosa, K., Brestic, M., Zivcak, M., Hossain, Z. (2014). Nano-CuO stress induced modulation of antioxidative defense and photosynthetic performance of syrian barley *(Hordeum vulgare L.)*. *Environ. Exp. Bot*. 102, 37-47.

Shaw, A. K. and Hossain, Z. (2013). Impact of nano-CuO stress on rice *(Oryza sativa L.)* seedlings. *Chemosphere*. 93, 906-915.

Sheykhbaglou, R., Sedghi, M., Tajbakhsh Shishevan, M., Seyed Sharifi, R. (2010). Effects of nano-iron oxide particles on agronomic traits of soybean. *Not. Sci. Biol*. 2, 112-113.

Simonin, M. and Richaume, A. (2015). Impact of engineered nanoparticles on the activity, abundance, and diversity of soil microbial communities: a review. *Environ. Sci. Pollut. Res*. 22, 13710-13723.

Singh, D. and Kumar, A. (2016). Impact of irrigation using water containing CuO and ZnO nanoparticles on Spinach oleracea grown in soil media. *Bull. Environ. Contam. Toxicol.* 97, 548-553.

Soni, D., Naoghare, P. K., Saravanadevi, S., Pandey, R. A. (2015). *Release, transport and toxicity of engineered nanoparticles: reviews of environmental contamination and toxicology.* D. M. Whitacre. Switzerland, Springer International Publishing.

Stampoulis, D., Sinha, S. K., White, J. C. (2009). Assay-dependent phytotoxicity of nanoparticles to plants. *Environ. Sci. Technol.* 43, 9473-9479.

Sun, T. Y., Gottschalk, F., Hungerbühler, K., Nowack, B. (2014). Comprehensive probabilistic modelling of environmental emissions of engineered nanomaterials. *Environmental Pollution.* 185, 69-76.

Taniguchi, N. (1996). *Nanotechnology: integrated processing systems for ultra-precision and ultrafine products.* Oxford University Press, USA.

Tripathi, S., Champagne, D., Tufenkji, N. (2012). Transport behavior of selected nanoparticles with different surface coatings in granular porous media coated with *Pseudomonas aeruginosa* biofilm. *Environ. Sci. Technol.* 46(13), 6942-6949.

Trujillo-Reyes, J., Majumdar, S., Botez, C. E., Peralta-Videa, J. R., Gardea-Torresdey, J. L. (2014). Exposure studies of coreeshell Fe/Fe_3O_4 and Cu/CuO NPs to lettuce *(Lactuca sativa)* plants: are they a potential physiological and nutritional hazard? *J. Hazard. Mater.* 267, 255-263.

Tolaymat, T., Badawy, A. E., Genaidy, A., Abdelraheem, W., Sequeira, R. (2017). Analysis of metallic and metal oxide nanomaterial environmental emissions. *J. Clean. Prod.* 143, 401-412.

Vallee, B. L., and Auld, D. S. (1990). Zinc coordination, function, and structure of zinc enzymes and other proteins. *Biochemist.* 29, 5647-5659.

Vance, M. E., Kuiken, T., Vejerano, E. P., McGinnis, S. P., Hochella, M. F., Rejeski, D., Hull, M. S. (2015). Nanotechnology in the real world: redeveloping the nanomaterial consumer products inventory. *Beilstein J Nanotechnol.* 6, 1769-80.

Wang, H., Kou, X., Pei, Z., Xiao, J. Q., Shan, X., Xing, B., (2011). Physiological effects of magnetite (Fe_3O_4) nanoparticles on perennial ryegrass *(Lolium perenne L.)* and pumpkin *(Cucurbita mixta)* plants. *Nanotoxicology*. 5, 30-42.

Wang, Z., Xie, X., Zhao, J., Liu, X., Feng, W., White, J. C., Xing, B. (2012). Xylem- and phloem-based transport of CuO nanoparticles in maize *(Zea mays L.)*. *Environ. Sci. Technol.* 46, 4434-4441.

Weir, A., Westerhoff, P., Fabricius, L., Hristovski, K., von Goetz, N. (2012). Titanium dioxide nanoparticles in food and personal care products. *Environ. Sci. Technol.* 46, 2242-2250.

Xu, C., Peng, C., Sun, L., Zhang, S., Huang, H., Chen, Y., Shi, J. (2015). Distinctive effects of TiO_2 and CuO nanoparticles on soil microbes and their community structures inflooded paddy soil. *Soil Biol. Biochem.* 86, 24-33.

Yadav, T., Mungray, A. A., Mungray, A. K. (2014). Fabricated nanoparticles: current status and potential phytotoxic threats. *Rev. Environ. Contam. Toxicol*. 230, 83-110.

Yang, Z., Chen, J., Dou, R., Gao, X., Mao, C., Wang, L. (2015). Assessment of the phytotoxicity of metal oxide nanoparticles on two crop plants, Maize *(Zea mays L.)* and Rice *(Oryza sativa L.)*. *Int. J. Environ. Res. Public Health*. 12, 15100-15109.

Ye, L., Li, L., Wang, L., Wang, S., Li, S., Du, J., et al., (2015). MPK3/MPK6 are involved in iron deficiency-induced ethylene production in Arabidopsis. *Front. Plant Sci.* 6:953.

Yoon, S. J., Kwak, J. I., Lee, W. M., Holden, P. A., An, Y. J. (2014). Zinc oxide nanoparticles delay soybean development: a standard soil microcosm study. *Ecotoxicol. Environ. Saf.* 100, 131-137.

You, T., Liu, D., Chen, J., Yang, Z., Dou, R., Gao, X., Wang, L. (2017). Effects of metal oxide nanoparticles on soil enzyme activities and bacterial communities in two different soil types. *J. Soils Sediments*. DOI 10.1007/s11368-017-1716-2.

Zafar, H., Ali, A., Zia, M. (2017). CuO nanoparticles inhibited root growth from *Brassica nigra* seedlings but induced root from stem and leaf explants. *Appl. Biochem. Biotechnol*. 181, 365-378.

Zhang, J., Zhang, Y., Chen, Y., Du, L., Zhang, B., Zhang, H., Liu, J., Wang, K. (2012). Preparation and characterization of novel polyethersulfone hybrid ultrafiltration membranes bending with modified halloysite nanotubes loaded with silver nanoparticles. *Ind. Eng. Chem Res*. 51, 3081-3090.

Zhao, L. J., Peralta-Videa, J. R., Ren, M. H., Varela-Ramirez, A., Li, C. Q., Hernandez-Viezcas, J. A., Aguilera, R. J., Gardea-Torresdey, J. L. (2012). Transport of Zn in a sandy loam soil treated with ZnO NPs and uptake by corn plants: electron microprobe and confocal microscopy studies. *Chem. Eng J*. 184, 1-8.

Zuverza-Mena, N., Medina-Velo, I. A., Barrios, A. C., Tan, W., Peralta-Videa, J. R., Gardea- Torresdey, J. L. (2015). Copper nanoparticles/ compounds impact agronomic and physiological parameters in cilantro *(Coriandrum sativum). Environ. Sci. Processes Impacts*. 17, 1783-1793.

Biographical Sketch

Tatiana Minkina

Affiliation: Southern Federal University

Research and Professional Experience: Full Professor, Doctor of Soil science and Doctor of Ecology

Professional Appointments: Head of Soil science and land resources evaluation department, Southern federal university

Honors:

Diploma of the Government of Rostov Region for the valiant achievements for the benefit of the Rostov region, Russia (2016).

Diploma of the Ministry of Education and Science of the Russian Federation for many years of long-term work for development and improvement of the educational process, significant contribution to the training of highly qualified specialists (2015).

Anniversary medal "Merits in the Southern Federal University" (2015).

The honorable certificate of achievements in the field of geochemistry (2013).

The honorable certificate "The best supervisor of students in Southern federal university" in a nomination "Natural sciences" (2010).

A bronze medal of the All-Russia exhibition center, the Russian agro-industrial week "Mellow autumn" of the Ministry of Agriculture of the Russian Federation "Working out of theoretical bases and innovative technological means of long-term preventive management of fertility biosystems" (2010).

Awarded by The Organizing Committee of 11th International Conference on the Biogeochemistry of Trace Elements, Florence, Italy (travel grant) (2011).

The honorable certificate "Best research scientific work in Southern Federal University" (2009).

Medal of the Society of soil scientists of V. V. Dokuchayev, Russia (2008).

Awarded by The 18th WCSS Organizing Committee, Philadelphia, Pennsylvania, USA, for participation in the Congress (2006).

Publications from the Last Three Years:

Over the past 3 years, 227 works have been published.

75 Documents, 247 total citations, h-index 10.

Web of Science: 62 Documents, 183 total citations, h-index 8.

ORCID ID 0000-0003-3022-0883

Scopus Author ID: 15063165400

Researcher ID: A-1683-2014

In: Soil Contamination
Editor: Jesper E. Lund

ISBN: 978-1-53613-266-3

Chapter 4

THE ASSESSMENT OF SOILS CONTAMINATION FROM U-238 AND U-234 DISEQUILIBRIA DATA

Jairo Roberto Jiménez-Rueda
and Daniel Marcos Bonotto*
Departamento de Petrologia e Metalogenia,
Universidade Estadual Paulista (UNESP),
Rio Claro, São Paulo, Brasil

ABSTRACT

Understanding processes involved in the soils formation is important for appropriate assessment of contamination occurring in the superficial horizons. Several tools have been used for such purpose, including the natural radionuclides. Uranium is a lithophile element that is among the main elements contributing to natural terrestrial radioactivity, together

* Corresponding Author Email: danielbonotto@yahoo.com.br.

with thorium and potassium (^{40}K). The two natural U-isotopes ^{238}U and ^{235}U are precursors of the natural mass number 4n+2 and 4n+3 decay series, respectively. The third natural U-isotope, ^{234}U, is radiogenic and generated in the ^{238}U decay chain according to the sequence: ^{238}U (4.46 Ga, α) $\rightarrow$ ^{234}Th (24.1 days, β^-) $\rightarrow$ ^{234}Pa (1.18 min, α) $\rightarrow$ ^{234}U (248 ka, α) $\rightarrow$ ... ^{206}Pb. If radioactive equilibrium occurs within the ^{238}U decay series, all radionuclides activities are equal. In closed systems, not subjected to weathering processes, ^{234}U reaches radioactive equilibrium with ^{238}U and the $^{234}U/^{238}U$ activity ratio (AR) is unity in the bulk of such systems. ^{234}U needs ~1.2 Ma to reach radioactive equilibrium with ^{238}U in closed systems consisting of many minerals and rocks. Equilibrium and disequilibrium conditions affecting ^{238}U and ^{234}U have been extensively used by the researchers since the 1960s for evaluating the time scale of processes controlling the formation and evolution of soils and weathering profiles. This chapter summarizes a study case focusing the presence of these natural U-isotopes in two soil profiles developed over sandstones from the Paraná sedimentary basin in São Paulo State, Brazil. The reported database has been used in order to evaluate the timescale of the weathering processes taking place there and also for suggesting implications of potential contamination sources into the different soils horizons.

INTRODUCTION

Weathering is an important phenomenon of the geochemical cycle as it contributes to the relief formation. It corresponds to a general term applied to physical and chemical changes suffered by rocks as a consequence of their exposition to different conditions of humidity and temperature (Bertin, 1974). Weathering is also considered a process related to rock disintegration or decomposition that is a common phenomenon occurring in the Earth's surface where the dominant factor is the rock interacting with water (Netto, 1980). In general, the weathering processes are slow and their type and duration are controlled by climate. Therefore, it is expected that the chemical weathering becomes accelerated under humid and hot climatic conditions.

When the soil formation takes place, there is a combination of effects of physical, chemical, biological, and anthropogenic processes on soil parent material. Thus, the soil genesis involves processes that develop layers or horizons in the soil profile, which cause additions, losses, transformations and translocations of material that compose the soil. Additionally, minerals derived from weathered rocks undergo further changes that cause the formation of secondary minerals and other compounds that are variably soluble in water. Such constituents are often moved from one area of the soil to others by water and animal activity so that the alteration and movement of materials within the soil causes the formation of distinctive soil horizons.

Cosmic-ray produced radionuclides (like ^{7}Be, ^{10}Be and ^{26}Al) coupled to geochemical mass balance methods have been recently used for investigating long-term denudation/erosion/weathering processes affecting hydrographic basins and soil formation (for instance, Baskaran, 1995; Small et al., 1999; Riebe et al., 2000, 2003, 2004).

Uranium is a lithophile element that contributes significantly to the natural terrestrial radioactivity. ^{238}U is the parent radionuclide of the (4n+2) mass number decay series. Some of its descendants have been used for studying weathering processes. The ^{238}U decay series finishes at the stable ^{206}Pb, according to the sequence (Figure 1): ^{238}U (4.49 Ga, α) $\rightarrow$ ^{234}Th (24.1 d, β^-) $\rightarrow$ ^{234}Pa (1.18 min, β^-) $\rightarrow$ ^{234}U (0.248 Ma, α) $\rightarrow$ ^{230}Th (75.2 ka, α) $\rightarrow$ ^{226}Ra (1622 a, α) $\rightarrow$ ^{222}Rn (3.83 d, α) $\rightarrow$ ^{218}Po (3.05 min, α) $\rightarrow$ ^{214}Pb (26.8 min, β^-) $\rightarrow$ ^{214}Bi (19.7 min, β^-) $\rightarrow$ ^{214}Po (0.16 ms, α) $\rightarrow$ ^{210}Pb (22.26 a, β^-) $\rightarrow$ ^{210}Bi (5 d, β^-) $\rightarrow$ ^{210}Po (138 d, α) $\rightarrow$ ^{206}Pb.

The radionuclides ^{238}U and ^{234}U have been used to evaluate the chemical weathering rate in hydrographic basins based on the fact that ^{234}U is preferentially mobilized to ^{238}U when rock weathers in the alteration front (Ivanovich and Harmon, 1992). Measurements of the $^{234}U/^{238}U$ activity ratio (AR) in rocks, soils and waters have allowed the calculation of the solution coefficient for the uranium characteristic of the region, that allows evaluate the time necessary to weather 1 m of rock under actual climatic conditions (Moreira-Nordemann, 1980, 1984; Conceição and Bonotto, 2003; Bonotto et al., 2007).

U-Isotopes Disequilibrium Theoretical Background

The two natural U-isotopes ^{238}U and ^{235}U are precursors of the natural mass number 4n+2 and 4n+3 decay series, respectively. The third natural U-isotope, ^{234}U, is radiogenic and generated in the ^{238}U decay chain as shown in Figure 1. The variation of the number of radionuclides per time unit in this series is:

$$
\begin{aligned}
dN_1 &= -\lambda_1 N_1 dt \\
dN_2 &= (\lambda_1 N_1 - \lambda_2 N_2)\, dt \\
dN_3 &= (\lambda_2 N_2 - \lambda_3 N_3)\, dt \\
&\ldots\ldots \\
dN_i &= (\lambda_{i-1} N_{i-1} - \lambda_i N_i)\, dt
\end{aligned} \qquad (1)
$$

where: N_i corresponds to the number of disintegrations associated to the *i-th* nuclide of the decay series in time t and λ_i is its respective decay constant. If the notation *"o"* represents the initial condition at time $t = 0$, on the assumption that $N_2^0 = 0$ and when $\lambda_2 >> \lambda_1$, the solution of the second equation from (1) provides the number of radionuclides produced by the radioactive decay of their parent, i.e.:

$$N_2 = \lambda_1 N_1^0 [1 - exp(-\lambda_2 t)] / \lambda_2 \qquad (2)$$

If the system exists for a long time ($t \rightarrow \infty$), then, eq. 2 may be simplified so that the number of atoms produced remains unaltered, allowing a re-write of eq. 1 to the situation named "Secular Radioactive Equilibrium" in which the number of atoms disintegrating per time unit is equal to that formed at the same time interval, also expressed as (Ivanovich and Harmon, 1992):

$$\lambda_1 N_1 = \lambda_2 N_2 = \lambda_3 N_3 = \ldots\ldots = \lambda_i N_i \qquad (3)$$

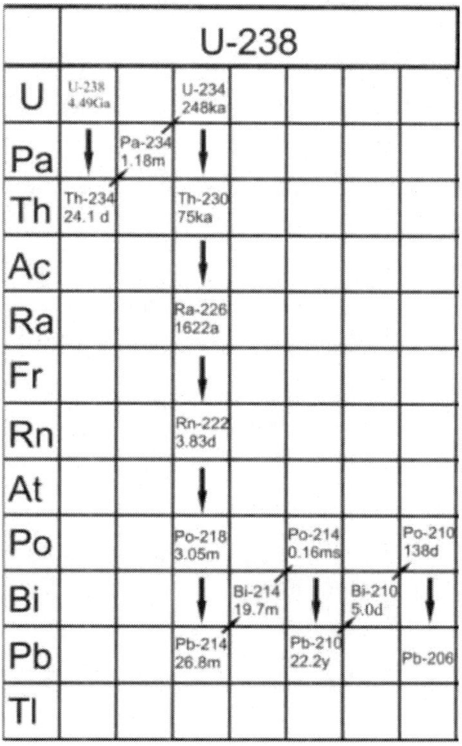

Figure 1. The natural uranium decay series. Modified from Ku (1976).

Thus, if radioactive equilibrium occurs within the ^{238}U decay series, the radionuclides activities are equal. In closed systems, not subjected to weathering processes, ^{234}U reaches radioactive equilibrium with ^{238}U and $\lambda_{238}N_{238} = \lambda_{234}N_{234}$ (the notation *"238"* refers to ^{238}U and *"234"* to ^{234}U). Therefore, the $^{234}U/^{238}U$ activity ratio (AR) is unity in the bulk of such systems. Because $\lambda_{238} << \lambda_{234}$, eq. 2 modified allows an estimate of the time for the establishment of the radioactive equilibrium between them, that is:

$$t = - \ln [1 - \lambda_{234}N_{234} / \lambda_{238}N_{238}] / \lambda_{234} \tag{4}$$

Equation 4 is unsolved under radioactive equilibrium conditions ($AR = 1$), but if $AR = 0.97$, then, $t = 3.5/\lambda_{234}$, implying the return to secular equilibrium over a timescale of circa 5 half-lives of the daughter nuclide. Thus, ^{234}U needs ~1.2 Ma to reach radioactive equilibrium with ^{238}U in closed systems consisting of many minerals and rocks. Latham and Schwarcz (1986, 1987), Ivanovich and Harmon (1992), and Suresh et al., (2013), among others, have discussed these concepts elsewhere.

Water/rock-soil interactions frequently result in AR's >1 for dissolved uranium and AR's <1 in the solid phase, mainly porous matrices (Osmond and Cowart, 1976; Ivanovich and Harmon, 1992). Alternative mechanisms may explain the disequilibria conditions affecting ^{238}U and ^{234}U in natural systems. For instance, Rosholt et al., (1963) proposed the occurrence of enhanced chemical solution of ^{234}U due to radiation damage of crystal lattices or to autoxidation from U^{4+} to U^{6+} on decay of the parent ^{238}U. Other significant early works in the 1960s demonstrated that a process of isotopic evolution of U and Th occurs in most types of sediments, altered volcanic ashes and deeply buried granites provided that some groundwater is allowed to migrate through the porous zones of these materials during their geologic history (Rosholt et al., 1966). Additionally, Hansen and Stout (1968) found that three soil profiles developed upon glacial till and loess contained excess ^{230}Th and quantities of ^{234}U slightly less than quantities of ^{238}U. Hansen and Stout (1968) proposed a model for migration of ^{238}U and ^{234}U in other soils, showing that ^{234}U was leached more readily than ^{238}U in the early stages of rock weathering, whereas some ^{234}U in excess of ^{238}U accumulated in the topsoils of two soils of high organic content. In the 1970s, the experiments performed by Kigoshi (1971) and Fleischer and Raabe (1978 a,b) showed that alpha-particle recoil ejection of the ^{234}U precursor, ^{234}Th, into solution may also generate enhanced $^{234}U/^{238}U$ activity ratios in the liquid phase.

Different approaches have been utilized for the application of U-series nuclides to the study of weathering processes like those reported by Latham and Schwarcz (1986, 1987), Ivanovich and Harmon (1992), Dequincey et al., (2002), Chabaux et al., (2003, 2006, 2008), Van Carlsteren and Thomas (2006), Bernal et al., (2006), Dosseto et al., (2008), etc. When the soils horizons exhibit AR's corresponding to unity, within experimental errors, they indicate radioactive equilibrium between ^{234}U and

^{238}U at least over the last 1 Ma (Latham and Schwarcz, 1987). Thus, there is no occurrence at these sites of extra ^{234}U-loss or ^{234}U-gain either due to enhanced chemical solution of ^{234}U due to radiation damage to crystal lattices or to autoxidation from U^{4+} to U^{6+} resulting from the decay of the parent ^{238}U (Rosholt et al., 1963; Fleischer, 1975) or by α-recoil effects (Kigoshi, 1971).

However, ^{234}U may be preferentially accumulated in some soil horizons, generating AR's > 1, within experimental errors, a typical result related to the effect of rock/soil-water interactions (Osmond and Cowart, 1976). In such cases, they provide useful timing information concerning their formation. These disequilibria allow an estimate of "apparent time" representing the state of disequilibrium acted against by the decay back tending to equilibrium of the shorter-lived daughter ^{234}U. This time (*t*) may be calculated from the initial ^{234}U-excess compared to the final ^{234}U equilibrium value (or close to it) according to the following equation (Latham and Schwarcz, 1986, 1987):

$$t = 1/\lambda \ln (D/P - 1) \qquad (5)$$

where: λ is the ^{234}U decay constant, and *D/P* is the activity of the daughter (*D*, ^{234}U) to its parent (*P*, ^{238}U) ($^{234}U/^{238}U$ in the present case).

Additionally, preferential removal of ^{234}U may occur at some sites, generating AR's<1, within experimental errors, that is also a typical result related to the effect of rock/soil-water interactions (Osmond and Cowart, 1976). Timing information may be generated as well, whilst the "apparent time" can be now estimated according to the expression (Latham and Schwarcz, 1986, 1987):

$$t = - 1/\lambda \ln (1 - D/P) \qquad (6)$$

In both cases, the time needed to restore the system to equilibrium from excess (eq. 5) or deficiency (eq. 6) may be calculated by the use of the quantity *5HF – t* (*HF* is the half-life of ^{234}U) (Latham and Schwarcz, 1986, 1987).

SOILS PROFILES DEVELOPED OVER SANDSTONES OF THE PARANÁ SEDIMENTARY BASIN

The Paraná sedimentary basin in South America, like others sedimentary sequences, has been extensively investigated due to its economic potential for supplying water, petroleum, coal, clays, and metals like uranium, among other resources. It is a giant geological unit, whose surface area corresponds to 1.6×10^6 km^2. It comprises southern Brazil (1×10^6 km^2 in the states of Mato Grosso, Mato Grosso do Sul, Goiás, Minas Gerais, São Paulo, Paraná, Santa Catarina and Rio Grande do Sul), eastern Paraguay (0.1×10^6 km^2), NW Uruguay (0.1×10^6 km^2) and the northeastern extreme corner of Argentina (0.4×10^6 km^2) (Figure 2) (Araújo et al., 1999). The basin is intercratonic, where the sedimentary sequence covering since the Silurian-Devonian up to the Cretaceous periods is almost undisturbed, with gentle dips towards the center of the basin.

Figure 2. Location of the Paraná sedimentary basin in South America.

Table 1. Brief horizons description of the two soil profiles sampled at São Paulo State, Brazil

Depth (m)	Horizon	Major Remark	Depth (m)	Horizon	Major Remark
Soil Profile 1 (P1) – Piracicaba city			**Soil Profile 2 (P2) – Rio Claro city**		
0.00-0.10	Ap	Superficial horizon	0.00-0.06	Ap	Superficial horizon
0.10-0.60	Bt_1	Saline concretions	0.06-0.15	AB	Very frequent thin roots
0.60-2.06	2Bticn	Coal presence	0.15-0.29	BA	Few thin roots
2.06-3.30	$Ctcn_1$	Soil pebbles	0.29-0.65	Bi	Large macro prismatic structure
3.30-4.62	3Bitcn	Coal presence	0.65-0.79	2Bicn	Goethite and hematite concretions
4.62-6.40	4Btcn	Light clayey	0.79-0.93	3Bicn2	Gleyzation
6.40-7.16	Ct_2	Soil pebbles	0.93-1.10	Fm-G	Goethite and hematite presence
6.40-7.16	Pedolith	Paleocutans	1.10-1.24	$4Fm\text{-}H_2$	Laterite cemented by hematite
7.16-7.54	Fm nodular	Laterites presence	1.50 (?)	Crf (PL-01)	Plintite
7.54-8.76	Crf nodular	Kaolinization	1.60 (?)	Crf (PL-02)	Plintite
8.76-9.74	Rrg	Kaolinization	outcrop	R	Sandstone from Pirambóia Formation
8.76-9.74	CrfBtg	Gleyzation			
8.76-9.74	Rrft	Kaolinization			
9.74-19.76	Crf	Plintite formation			
9.74-19.76	Rrf	Plintite formation			
19.76-20.50	R	Sandstone from Tatuí Formation			

Sampling of the soil profiles considered in this study was done close to Piracicaba and Rio Claro cities, almost in the center of São Paulo State, Brazil. The soil profile 1 (P1) was situated about 8 km distant from Piracicaba city by the SP-304 highway that connects Piracicaba and Santa Bárbara d'Oeste cities. The soil profile 2 (P2) was situated circa 10 km distant from Rio Claro city by an unpaved road giving access to sugar cane plantations in the region that is connected to SP-191 highway between Rio Claro and Araras cities.

The soils horizons from P1 developed over Permian sediments of Tatuí Formation (Daemon and Quadros, 1970) whose lithology derives from sediments of the post-glaciation period (Itararé Group). Tatuí Formation is mainly characterized by siltstones (Gimenez, 1996), despite also occurs layers of sandstones, limestones, shales and silex (Almeida and Melo, 1981). The soils horizons from P2 developed over Pirambóia Formation that is a sedimentary unit of expressive occurrence in central eastern of São Paulo State, whose deposition period was between the Lower Triassic and Upper Jurassic (Almeida and Melo, 1981). The Lower Member of Pirambóia Formation presents clayey facies of plane-parallel and crossed slotted stratification, whereas the Upper Member is characterized by the presence of little to very clayey sandstone benches with plane-tangential crossed stratification of small to mean size. Some major features of both soil profiles P1 and P2 are summarized in Table 1.

WEATHERING INDICES AND AGES

Several indexes have been reported in the literature for expressing the degree of chemical weathering of a mineral or soil material and some of them will be considered in this chapter. Ratios of moles of different constituents of rocks, minerals and soils have been employed to condense elemental analyses of these materials, for instance (Jackson and Sherman, 1953): silica:alumina ratio (SiO_2/Al_2O_3), silica:ferric oxide ratio (SiO_2/Fe_2O_3), silica:sesquioxide ratio ($SiO_2/Al_2O_3+Fe_2O_3$), bases:alumina ratio (K_2O+Na_2O+CaO/Al_2O_3), etc.

Alumina has often been employed as the basis for calculation of relative weathering losses of other constituents and, hence, the referred ratios serve as measures of depletion of the other constituents. In many soil profiles, ratios involving alumina are calculated considering the leached horizon and the parent material.

The Fe_2O_3:FeO ratio is a measure of the oxidation that occurs during weathering, which is particularly important for weathering rinds as they are defined largely by color. Fe_2O_3 is generally stable in most weathering environments (Loughnan, 1969) but any depletion of Fe_2O_3 and the potential mobility of FeO may affect the Fe_2O_3:FeO ratio as a measure of oxidation.

Parker (1970) proposed a weathering index in order to measure the loss of bases during weathering. The index is not referenced to the sesquioxide (R_2O_3) ratio or to any other stable constituent, so different samples (that have different initial compositions) are not directly comparable (Colman, 1982). However, the index changes rapidly as bases are lost, so it is a sensitive indicator of early stages of weathering in a given sample (Colman, 1982). The index is defined as:

$$P = 100\ (K/0.25 + Na/0.35 + Ca/0.7 + Mg/0.9) \quad (7)$$

where the denominators represent measures of the strengths of the cation-oxygen bond.

Reiche (1943, 1950) devised two indices that incorporate almost all of the major elements involved in weathering. They are based on molecular percentages and are called the weathering potential index (WPI) and the product index (PI), which are defined as:

$$WPI = 100\ (\Sigma bases - H_2O) / (\Sigma bases + SiO_2 + \Sigma R_2O_3) \quad (8)$$

$$PI = 100\ (SiO_2) / (SiO_2 + \Sigma R_2O_3) \quad (9)$$

Because the WPI-PI plot combines all the major elements involved in weathering (bases, silica, R_2O_3, and water), it has been considered an optimum two-dimensional portrayal of chemical weathering (Colman,

1982). However, Colman (1982) also pointed out that it has the disadvantage of not showing changes in individual elements, and because it is based on percentage data, it does not account for compensating (relative) changes in elements or groups of elements.

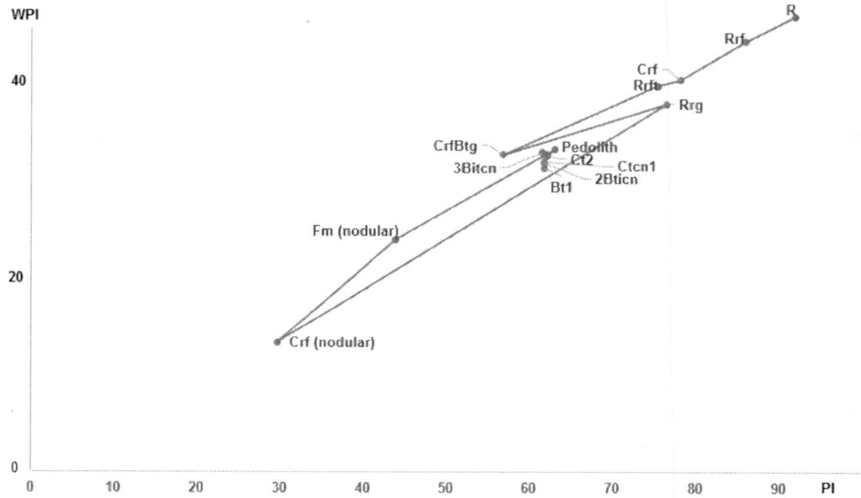

Figure 3. WPI *vs.* PI plot for soils horizons from profile P1, Piracicaba city, São Paulo State, Brazil.

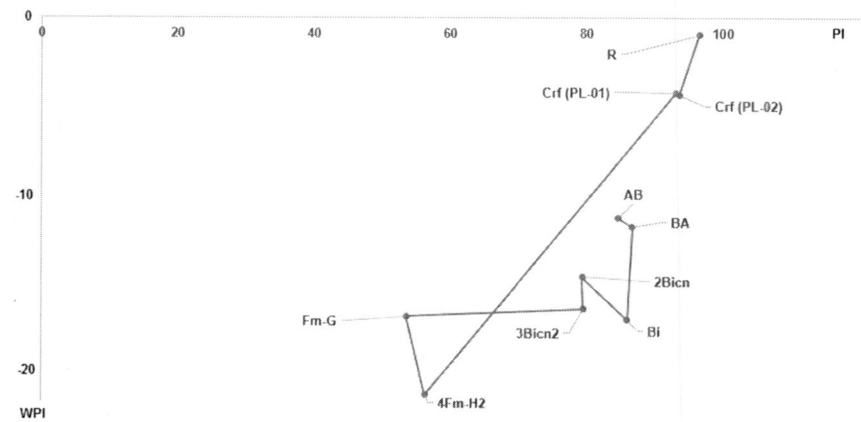

Figure 4. WPI *vs.* PI plot for soils horizons from profile P2, Rio Claro city, São Paulo State, Brazil.

Table 2. Results of the chemical analyses in the soils profiles P1 and P2 collected at São Paulo State, Brazil. All values are given in wt. percentage (%)

Horizon	SiO_2	Na_2O	K_2O	CaO	MgO	Al_2O_3	P_2O_5	Fe_2O_3	MnO	TiO_2	LOI
Soil Profile 1 (P1) – Piracicaba city											
Bt_1	55.32	0.07	0.05	0.04	0.03	19.11	0.05	12.99	0.05	2.22	10.06
2Bticn	55.77	0.07	0.05	0.03	0.06	19.59	0.06	12.67	0.05	2.26	9.40
$Ctcn_1$	55.88	0.07	0.04	0.03	0.05	19.55	0.06	12.89	0.05	2.29	9.13
3Bitcn	56.54	0.07	0.05	0.04	0.05	19.89	0.05	13.15	0.05	2.38	7.74
4Btcn	56.71	0.07	0.05	0.05	0.03	19.36	0.05	13.08	0.05	2.27	8.31
Ct_2	56.62	0.07	0.05	0.04	0.03	19.50	0.05	12.68	0.05	2.36	8.55
Pedolith	57.65	0.07	0.05	0.04	0.02	19.58	0.05	11.90	0.05	2.42	8.19
Fm nodular	39.44	0.05	0.67	0.03	0.27	16.54	0.14	32.11	0.02	1.74	8.97
Crf nodular	26.20	0.02	0.16	0.04	0.11	14.63	0.27	46.26	0.01	1.23	11.08
Rrg	70.89	0.08	0.78	0.04	0.26	13.90	0.14	6.70	0.01	1.24	9.45
CrfBtg	50.05	0.08	1.55	0.03	0.65	23.92	0.10	12.69	0.05	1.44	5.96
Rrft	69.53	0.08	1.13	0.04	0.39	14.31	0.14	7.46	0.01	0.98	5.94
Crf	71.79	0.09	1.09	0.04	0.37	17.06	0.04	2.25	0.01	0.91	6.37
Rrf	81.77	0.08	0.46	0.05	0.15	6.68	0.36	6.15	0.01	0.66	3.63
R	88.04	0.13	1.12	0.07	0.40	6.74	0.02	0.61	0.01	0.37	2.51

Table 2. (Continued)

Horizon	SiO_2	Na_2O	K_2O	CaO	MgO	Al_2O_3	P_2O_5	Fe_2O_3	MnO	TiO_2	LOI
Soil Profile 2 (P2) – Rio Claro city											
AB	65.99	0.04	0.03	0.03	0.06	12.47	0.06	9.70	0.03	2.69	8.90
BA	68.22	0.05	0.03	0.03	0.05	11.18	0.05	8.51	0.02	2.47	9.39
Bi	63.61	0.05	0.03	0.03	0.07	11.82	0.05	9.00	0.02	2.58	12.74
2Bicn	56.70	0.04	0.03	0.04	0.07	12.73	0.06	17.10	0.03	2.60	10.59
3Bicn2	55.55	0.04	0.04	0.04	0.09	12.06	0.05	17.86	0.02	2.60	11.65
Fm-G	31.92	0.03	0.04	0.03	0.07	14.18	0.09	41.11	0.01	2.02	10.23
4Fm-H_2	31.92	0.04	0.03	0.03	0.06	10.92	0.07	42.80	0.02	1.89	12.22
Crf (PL-01)	83.73	0.13	0.05	0.04	0.10	9.13	0.02	2.19	<0.01	0.48	4.14
Crf (PL-02)	84.72	0.11	0.06	0.04	0.08	8.52	0.02	1.71	<0.01	0.52	4.21
R	91.15	0.25	0.74	0.07	0.08	5.29	0.02	0.54	0.01	0.05	1.81

Table 3. U-isotopes data and ages estimated for developing horizons in the soil profiles P1 and P2 sampled at São Paulo State, Brazil

Soil Profile 1 – Piracicaba city			**Soil Profile 2 – Rio Claro city**		
Horizon	$^{234}U/^{238}U$ A.R.[1]	Time (Ma)	Horizon	$^{234}U/^{238}U$ A.R.[1]	Time (Ma)
Bt_1	0.99 ± 0.18	1.24	AB	1.05 ± 0.06	0.17
2Bticn	1.19 ± 0.16	0.64	BA	1.33 ± 0.20	0.84
$Ctcn_1$	2.40 ± 0.36	3.39	Bi	1.13 ± 0.07	0.51
3Bitcn	1.40 ± 0.18	0.91	2Bicn	1.21 ± 0.09	0.68
4Btcn	1.36 ± 0.23	0.87	3Bicn2	1.36 ± 0.14	0.87
Ct_2	1.32 ± 0.11	0.83	Fm-G	1.09 ± 0.14	0.38
Pedolith	1.18 ± 0.10	0.63	$4Fm\text{-}H_2$	1.15 ± 0.09	0.56
Fm nodular	1.08 ± 0.19	1.24	Crf (PL-01)	nc	nc
Crf nodular	0.80 ± 0.06	0.66	Crf (PL-02)	nc	nc
Rrg	2.50 ± 0.67	3.47	R	nc	nc
CrfBtg	2.74 ± 0.60	3.61			
Rrft	1.48 ± 0.06	0.98			
Crf	3.01 ± 1.00	3.72			
Rrf	2.52 ± 0.91	3.49			
R	2.03 ± 0.42	2.48			

[1]Analytical uncertainty corresponding to 1σ standard deviation; nc = not computed as the U content was below the detection limit (0.02 µg/g).

Table 2 reports the results of the chemical analyses in the soils profiles P1 and P2 as done by X-ray fluorescence analysis (XRF). LOI (Loss on Ignition) expresses the organic matter + adsorbed water + water in crystal lattices and fluid inclusions + CO_2 of carbonates + SO_2 of sulfides. SiO_2 is the dominant constituent in practically all horizons, except in *Crf nodular* at profile P1 where Fe_2O_3 is the major oxide. Diagrams of the WPI index against the PI index are plotted in Figure 3 and 4 from data reported in Table 2 for soils profiles P1 and P2.

The timing information for soils profiles P1 and P2 was generated considering eqs. 5 and 6, depending if the AR's values were higher or lower than unity, respectively. Table 3 shows the ages obtained in the calculations from the AR's data acquired by alpha spectrometry according to the analytical protocols described by Bonotto (2010). In some horizons from Profile 1, when the AR's values were 2 or higher, additional 1.24 Ma has been added in order to take into account each step needed for reaching successive radioactive equilibrium conditions. Such approach is different of that adopted by Bonotto et al., (2017), thus, explaining the ages higher than 1.24 Ma in some horizons of profile P1 as reported in Table 3.

Evolutionary Trends from Chemical and U-Isotopes Data

The R volume of the soil profile developed over sandstones of Tatuí Formation exhibits reddish colors, representing the occurrence of oxidation processes in reduced media from the original grayish rocks (Jackson and Sherman, 1953). Thus, ancient grayish clays result in stained/variegated of intense reddish up to reddish (rust) brown colors inside of a dominant greenish/grayish mass (matrix), manifesting in a reticulated form as a result of the structural regional conditioning present at the area. The fracture systems favor the secondary mineral formation such as mineral clays associated or not with other types of organic/mineral colloids fulfilling these fractures.

These phenomena of new minerals formation and colloids accumulation favor the transport of surface waters plenty of oxygen, carbon, humates and other elements, giving origin to sediments with the presence of trees in swamps or flooded areas, such as the silt-sandy (and sometimes conglomeratic) Tatuí Formation. This redox process, named Plintification, characterizes a high topographic synclinal resulting from a sequential uplift of a zone that has suffered epirogenic modifications of variable intensity and frequency in different geologic periods.

They are also called Neotectonic reactivations, whose development was mostly during the Upper Miocene up to the Inferior-Medium Pliocene, when the climatic conditions should be sub-humid to humid. Consequently, the profile shows a silica and earth-alkaline elements remobilization that promotes a 2:1 type (*bissialitic)* plasma reorganization and typical plintification of structural lows and topographic highs showing iron oxides precipitations in an ocher/leucocratic matrix covered by a deep alitization mantle (laterites) or mature alitization (latossolization). This last process presents a volume aggregate of weathered covers/diagnostic horizons constituting the lateritic profile, which exhibits (sometimes decametric) latosolic/oxic horizons followed by lateritic/petroplintites horizons, stained/variegated plintites horizons, *monosialitic* or *bissialitic* horizons, which depend on structures favoring reactions developed under tropical/subtropical climatic conditions.

In terms of timing perspectives, it may occur a chronologic continuity (Chronos) or discontinuity (Kairos), which often manifest as colloidal or mineral neo/retroconcentrations, thus, generating a retrochronology by the neomineralization effects in depth that can correlate the epipedogenic volumes/horizons with the same evolutive age. This permits the environment identification in which such interaction developed and idealize the paleoenvironmental reconstruction of landscapes/soils. These horizons can be metrics yet, resting over a preexistent rocky mantle. In this case, they occur in a region affected by several morphoclimatic cycles where tectonic activities and climatic changes acted as designer/modelling factors of the erosive processes generation. They imply on more recent age (2.48 Ma) for the regolith and saprolite lithogenic mantles described as

Plintites *(monosialitic/bissialitic),* iron-clayey or ferri-*argilans,* which developed in the Upper Miocene/Lower-Medium Pliocene.

In that period, the morphoclimatic environment tended to be more subtropical, contributing with the bissialitization in depth as a final effect of a relatively transitional period, when happened a Kairos event with reactivations promoting variations of the morphoclimate conditioners. This characterized a tropical humid climate as evidenced by the intense alitization processes and also by the strong variegated brown reddish colors that generated a Rrf and Crf horizon volume identified as pale reddish/light red. The silty rock matrix is still preserved despite it has been weathered from 3.72 up to 3.49 Ma under humid tropical climatic conditions.

Later, during the Lower-Medium Pleistocene, there is generation of records marked by neoformation and translocation of clay minerals among fractures series plans. It becomes evident volumes/horizons with presence of cutans/ferri-*argilans* films and verticalized clayey goethans that sometimes form red to brown reddish reticulated systems inserted in an ocher (greenish/gray-greenish) matrix. It may have developed between 980 and 910 ka, whilst the migrations and superimposed cutans neoformations possibly were favored by weathering processes under sub humid/dry warm tropical conditions, generating the 8-9 m deep Rrft horizon (980 ka) and horizons 3Bitcn2 (910 ka), 4Btcn3 and Ct2 (830 ka).

Therefore, the transitional morphoclimatic interaction effects before a very wet and hot tropical climate are better defined, promoting the generation of ferruginous nodules characterizing the alitization origins developed from 900 to 830 ka. They caused overlapped neo-feedbacks weathering records still with domain of timeless intervals of higher transforming intensity when generating ferri-*argilans* and limonitic iron reconcentrations between fractures joints or fractures series. They exhibit botryoidal/nipple shapes of variable size (5-10 cm) and are named here concretionary despite represent nodules of *in situ* formation.

During the Lower Pleistocene (660-630 ka), there are records of joint monosialitization, bissialitization and alitization, characterizing processes developed under morphoclimatic conditioning varying between subtropical-(sub)dry and very humid-hot tropical, which promoted the formation of a decimetric Fm volume (nodular laterite/nodular petroplintite) at 7.50 m depth. It has developed about 660 ka, associated to pedolith originated by intra lithogenic transports within cracks or fracture cavities originated by oxidation processes as a consequence of the surface waters circulation that are saturated with oxygen and anhydrous carbon (Jackson and Sherman, 1953). Therefore, they contribute with the oxidation process and joint alitization/monosialitization occurring in the 2Btcn (2Bitcn) volume (640 ka) under very wet and hot tropical conditions (~ 400 ka). There are found records of the Crf volume that presents a more intense plintification characterizing the effects of a very hot tropical climate with high rainfall.

Additionally, during the Medium Pleistocene (640 ka), there are records of plasma (colloidal) reorganization generated under morphoclimatic (sub)humid/(sub)dry subtropical conditions, developing the 2Btcn volume (251 ka). However, in the Upper Pleistocene (65.8-32.8 ka), the monosialitic and *bissialitic* volumes are again dominant under (sub) humid and (sub) dry subtropical conditions. Therefore, the Profile 1 represents the evolutionary processes occurring at the Piracicaba city region since the Upper Miocene to the Lower Pliocene, in which there are records of factors and typical processes acting in the very wet and hot tropical morphoclimatic domain during which developed the volumes of weathered covers characterized as Crfcn (nodular) (3.72 Ma). As a consequence, it remains a record in silt-clayey materials of the Tatuí Formation evidencing variegated *latossolization* and laterization actions that typify the volumes of plintites and the monosialitic domain characterized as Crf (also 3.72 Ma).

Thus, the beginning of the morphoclimatic changes under (sub) dry/(sub) wet subtropical conditions and with deformative subsidence effects conditionate a *graben* allowing the development of anaerobic conditions that induce the generation of *smectite* clay minerals or overlap mono-bissialitization interstratified processes characterized as CrfBtg (age 3.61 Ma). This generates two other records (one at 3.47 Ma), originating the Rrg volume that confirms its anaerobic situation extending up to the start of a new morphoclimatic change due to the effects of the block raising. After returning to the *horst* position under subtropical/tropical humid morphoclimatic action, the generation of ferric nodules masking/hiding the paleocutans and neoforming ferri-*argilans* indicate an absolute age of Medium Pliocene (3.39 Ma). In this time, returns the morphoclimatic interferences with the generation of a *graben* that promotes a domain of bissialitization processes in anaerobic conditions between 2.48 Ma (Rtg volume) and 980 ka when is generated the Rrft volume under subtropical – tropical climate. Then, plintification occurs evidencing latossolization and mono-bissialitization processes characterizing a 3Bitcn volume that indicates an absolute age of 910 ka under (sub)humid subtropical conditions in *graben* positioning during the Plio-Pleistocene.

During the Lower Pleistocene and under (sub)wet/(sub)dry tropical conditions were generated the weathered covers 4Btcn3 (870 ka) and Ct (830 ka). The Fm volume developed in the Medium Pleistocene (660 ka), characterizing a typical action under very humid and hot tropical conditions extending until the generation of the 2Bticn volume (640 ka) and pedoliths (soils additions by slow mass movements between layers and lens) (630 ka). The records of a Bt1 volume developed more recently, characterizing monosialitic argillation processes with the presence of ferri-*argilans* cutans that indicate the morphoclimatic conditioning of structural lows/topographic highs situations under humid/(sub)dry subtropical climatic conditions that possibly extended up to 65.8 ka.

CONCLUSION

The soils considered in this study are dark red latossols according to the Brazilian System of Soils Classification. Thus, in these types of crusts/weathered covers, the elements may migrate in the colloids form, originating the sols (colloidal solutions) and causing particularly common migration in regions characterized by hot and wet climates, where the acid waters are enriched in organic substances. After acquiring mobility, the colloidal minerals migrate in soils and weathered crusts from wet regions in the form of dispersions. Their particles become orientated during the transport and deposition, developing metacolloids and micro-colloidal structures that congregate several receptors characterized by the presence of overlapped films indicating neo-reweathering effects typical of interstratified colloids and clays (cutans) that are situated on the walls cavities and fissures (plasma films). Their geochemical importance is not very clear but they take an important role as sorption barriers slowing the mobility of some contaminants that may affect the quality of surface and groundwaters. Thus, soils covers in humid and hot tropical or subtropical regions affected by uplifted (faulted) structural lows can originate alternate *horsts* and *grabens*. Such blocks may favor the development of colloidal forms and occurrence of adsorption phenomena. Additionally, they cause the contaminants absorption by organic colloids or fine particles of the silt/clay grain size, which difficult their removal. These and other conditioning structures form reactions and plasma reorganizations resulting from alteration processes and neoformation of silicate minerals associated to colloidal organo-minerals present in profiles named lateritic latossols. They may develop in structurally appropriate regions for the installation of wastes disposal areas like landfills, sewage treatment, cemeteries and industrial wastes reuse, for example, from sugar cane crops, where potassium can be used as fertilizer. Additionally, they can be also utilized for avoiding the surface and groundwater contamination, thus, permitting a secure supply of this important resource to the actual and future communities. When their genetic lithologies comprise clays or minerals potentially able to weather into clays along the time actions, then, the local, regional or continental structures can be truly natural sites mitigators to control and guide the industrial spoils stock up to the development of new technologies aimed to their reuse and safe application.

ACKNOWLEDGMENTS

FAPESP (Foundation Supporting Research in São Paulo State) and CNPq (National Council for Scientific and Technologic Development) in

Brazil are greatly thanked for financial support of related investigations described in this chapter.

References

Almeida, F. F. M. de; Melo, M. S. *In Geological map of the State of São Paulo*; Pro-Ore; Ed; PROMOCET: São Paulo, 1981; v. 1, pp. 46-81.

Araújo, L. M.; França, A. B.; Potter, P. E. *Hydrogeol J* 1999, 7, 317-336.

Baskaran, M. *J Geophys Res* 1995, 100, 2833-2840.

Bernal, J. P.; Eggins, S. M.; McCulloch, M. T.; Grün, R.; Eggleton, R. A. *Chem Geol* 2006, 235, 76-94.

Bertin, L. *The new Larousse Encyclopedia of the Earth*; Crown Publishers, Inc.: New York, 1974; 425 pp.

Bonotto, D. M. The Poços de Caldas *Hot Spot: A Big Blast for Nuclear Energy in Brazil*; Nova Science Publishers, Inc: Hauppauge, New York, 2010.

Bonotto, D. M.; Fujimori, K.; Moreira-Nordemann, L. M. *Appl Radiat Isot* 2007, 65, 474-481.

Bonotto, D. M.; Jiménez-Rueda, J. R.; Fagundes, I. C.; Fonseca Filho, C. R. A. *Appl Radiat Isot* 2017, 119, 6-15.

Chabaux, F.; Dequincey, O.; Levesque, J. J.; Leprun, J. C.; Clauer, N.; Riotte, J.; Paquet, H. *CR Geoscience* 2003, 335, 1219-1231.

Chabaux, F.; Pelt, E.; Lucas, Y.; Allard, T.; Fritsch, E.; Balan, E.; Selo, M. *Geochim Cosmochim Acta* 2006, 70 S1, A93.

Chabaux, F.; Bourdon, B.; Riotte, J. In *Radioactivity in the Environment: U-Th Series nuclides in aquatic systems*; Krishnaswami, S.; Cochran, J. K.; Eds.; 2008; v. 13; pp. 49-104.

Colman, S. M. *US Geol Surv Prof Paper* 1982, 1246, 1-51.

Conceição, F. T.; Bonotto, D. M. *Environ Geol* 2003, 44, 408-418.

Daemon, R. F.; Quadros, L. P. *In Proceedings of the 24th Brazilian Congress of Geology*; SBG-Brazilian Society of Geology: Brasília, 1970; pp. 359-412.

Dequincey, O.; Chabaux, F.; Clauer, N.; Sigmarsson, O.; Liewig, N.; Leprun, J. C. *Geochim Cosmochim Acta* 2002, 66, 1197-1210.

Dosseto, A.; Turner, S. P.; Chappell, J. *Earth Planet Sci. Lett* 2008, 274, 359-371.

Fleischer, R. L. *Health Phys* 1975, 29, 69-73.

Fleischer, R. L.; Raabe, O. G. *Geochim Cosmochim Acta* 1978a, 42, 973-978.

Fleischer, R. L.; Raabe, O. G. *Health Phys* 1978b, 35, 545-548.

Gimenez, N. L. B. *Petrological study of sandstones from Tatuí Formation at São Paulo State*; Ms. Dissertation; UNESP-São Paulo State University: Rio Claro, SP, 1996; 140 pp.

Hansen, R. O.; Stout, P. R. *Soil Sci* 1968, 105, 44-50.

Ivanovich, M.; Harmon, R. S., *Uranium Series Disequilibrium: Applications to Environmental Problems*; Clarendon Press: Oxford, 1992; 2nd ed.

Jackson, M. L.; Sherman, G. D. *Adv in Agron* 1953, 5, 219-318.

Kigoshi, K. *Science* 1971, 173, 47-48.

Ku, T. L. *An Rev Earth Plan Sci* 1976, 4, 347-379.

Latham, A. G.; Schwarcz, H. P. 1986. Models of uranium etching and leaching: their applicability to estimating rates of natural uranium removal from crystalline igneous rocks using U-series disequilibrium data from the Eye-Dashwa Lakes pluton; *Canadian Atomic Energy*, Rep. TR-353, 1986; 60 pp.

Latham, A. G.; Schwarcz, H. P. *Appl Geochem* 1987, 2, 55-65.

Loughnan, F. C. *Chemical weathering of the silicate minerals*; Elsevier: New York, 1969; 154 pp.

Moreira-Nordemann, L. M. *Geochim Cosmochim Acta* 1980, 44, 103-108.

Moreira-Nordemann, L. M. *J Hydrol* 1984, 71, 131-147.

Netto, A. S. T. *Sedimentology Manual*; Petrobrás/SEPES/DIVEN: Salvador, 1980; 194 pp.

Osmond, J. K.; Cowart, J. B. *At Energy Rev* 1976, 14, 621-679.

Parker, A. *Geol Mag* 1970, 107, 501-504.

Reiche, P. *J Sediment Petrol 1943*, 13, 58-68.

Reiche, P. *A survey of weathering processes and products*; New Mexico University Publication in Geology, 1950; 95 pp.

Riebe, C. S.; Kirchner, J. W.; Granger, D. F.; Finkel, R. C. *Geology* 2000, 28, 803-806.

Riebe, C. S.; Kirchner, J. W.; Finkel, R. C. *Geochim Cosmochim Acta* 2003, 67, 4411-4427.

Riebe, C. S.; Kirchner, J. W.; Finkel, R. C. *Earth Planet Sci Lett* 2004, 224, 547-562.

Rosholt, J. N.; Shields W. R.; Garner, E. L. *Science* 1963, 139, 224-226.

Rosholt, J. N.; Doe, B.; Tatsumoto, M. *Geol Soc Am Bull* 1966, 77, 987-1004.

Small, E. F.; Anderson, R. S.; Hancock, G. S. *Geomorphology* 1999, 27, 131-150.

Suresh, P. O.; Dosseto, A.; Hesse, P. P.; Handley, H. K. *Earth Planet Sci Lett* 2013, 379, 26-37.

Van Carlsteren P.; Thomas, L. *Earth-Sci. Rev 2006*, 75, 155-175.

In: Soil Contamination
Editor: Jesper E. Lund

ISBN: 978-1-53613-266-3

Chapter 5

CLEAN-BIOTECHNOLOGY WITH ENHANCED PLANT-MICROBIAL SYNERGISTIC ACTION: A NOVEL AND FUNCTIONAL REMEDIATION MECHANISM FOR PESTICIDE POLLUTED SOILS

***M. D. Asemoloye*[1,*]*, O. M. Olowe*[2] *and S. G. Jonathan*[1]**
[1]Department of Botany, University of Ibadan,
Ibadan, Oyo State, Nigeria
[2]Department of Pure and Applied Botany, College of Biological Sciences, Federal University of Agriculture,
Abeokuta, Ogun State, Nigeria

[*] Corresponding Author Email: asemoloyemike@gmail.com

ABSTRACT

Pesticides are generally used for the control of pests and weeds, they are of immense importance and cannot be eradicated, their abuse however may become an issue of concern. Many banned pesticides are still in use under different trade names especially in developing countries due to their active and faster actions and many of such pesticides are grouped as Persistent Organic Pollutants (POPs). POPs usually accumulate in the environment and become persistent with concurrent harmful health hazards. More also, many persistent pesticides are biogenetic in nature posing negative effects on wide range of living organisms at different trophic levels of an ecosystem. Generally, several remediation methods have been suggested for removal of persistent pesticides in the environment 'Bioremediation' (the use of living organisms) is currently gaining attention of biologists and environmental scientists, due to its environment friendly, applicability and cost effective approaches. The use of plants have been well studied for effective remediation of pesticides in soils (Phytoremediation), however, this Bioremediation mechanism is still facing problems such as phytoextraction and phytoaccumulation leading to accumulation of toxic pesticide residues in plant tissues. Plants as primary producers in the ecosystem may transfer toxins into primary consumers then to the secondary consumers, on and on across different food chain. Another problem of Phytoremediation is the difficulty in establishment of plant's growth in heavily polluted soil. In this chapter, synergistic biological mechanisms were suggested as safe, less expensive, environment friendly method for remediation of pesticide polluted soils. We are drawing attention to the use of synergistic plant-microbial mechanism as an effective and speedy method of remediation of pesticide-polluted soils feasible at large and small scale. In addition, we took insights into the roles played by active enzymatic secretions by soil microorganisms and their genetic inferences during synergistic pesticide degradation. Plant-microbial mechanism relies on symbiotic association between plant root and soil microorganisms leading to exchange of benefits for sustainable adaptation and survival in heavily polluted soils. It also promotes biodegradation and mineralization of pesticides in soil over phytoextraction and accumulation. Many plants and microorganisms can therefore be screened for their synergistic potentials in pesticides removal/degradation in the soil and this is possible with good knowledge of the particular pesticide's nature, its chemical mechanisms, its reactivity combined with adequate knowledge of the plant and microbial genetic and biological nature. These factors are necessary to set-up a sustainable synergistic clean-biotechnology for remediation of a pesticide polluted soil which will be feasible in large and small scale.

Keywords: persistent, biogenic, ecosystem, synergistic, bioremediation, clean-technology

1. Introduction

Today, the complexity in anthropogenic activities has severely affected the natural planetary design and exchange of materials between different spheres of lives such as climate, gaseous exchange, nutrient/water recycling, biogeochemicals cycle, food chain, and preservation of biodiversity or natural resources (Teng et al., 2012; Alloway and Trevors, 2013; Valentine et al., 2013). Estimates have shown that about 30% of the human environment is impacted or contaminated with different pollutants including contemporary pollutants such as heavy metals, hydrocarbons and pesticides (Alloway and Trevors, 2013; Valentín et al., 2013), many pesticides such as perfluorooctanoic acid (PFOA), polybrominated diphenyl ethers (PBDEs) and polychlorinated naphthalenes (PCNs), have been reported with capacity to evolve into new generation persistent organic pollutants (POPs) (Lohmann et al., 2007). Generally, these POPs are grouped into Organophosphate and Organochline pesticides (Asemoloye et al., 2017a). Today, there are many reported mechanisms for removal of persistent pesticides in the environment due to their injurious health effects. Methods suggested include mechanical, chemical, physical and physico-chemical processes, several researchers however reported 'Bioremediation mechanisms' (the use of living organisms) as an environmentally—safe and less energy demanding mechanism. This mechanism has advantage over physic-chemical or mechanical mechanisms, which are very effective but usually energy/capital intensive, intrusive to the environment and difficult to establish in small scale remediation (Segura and Ramos, 2013; Asemoloye et al., 2017 b&c). Bioremediation mechanism utilizes living organisms, their part or products to remediate, degrade or mineralize inorganic/organic contaminants from the human environment. Bioremediation mechanism in different forms, the use of plants for example is termed 'Phytoremediation', bacteria and fungi 'Microremediation', algae 'Phycoremediation', nematodes and so on.

Bioremediation is a promising mechanism because of its relatively low environmental disturbance, cost and it is highly acceptable by the public compared with the other conventional methods. As promising as Bioremediation mechanism is, there are problems it is facing, which include; the challenges of establishing living organisms in heavily polluted site due to unavailability of nutrients, altered environmental conditions due to pesticides reactions and complexes formed by the POPs. It is therefore necessary to think about an achievable method to solve or alleviate such problems particularly when setting up this mechanism. Phytoremediation as well as Microremediation are known to be an environment-friendly and cost effective which were established to provide several benefits for the soil ecosystem in terms of carbon sequestration, soil quality, biomass, biofuel, and maintenance of soil biodiversity (Rajkumar et al., 2012). It is then necessary to circumvent these bioremediation problems by combining different bioremediation mechanisms synergistically. Legumes for example are essential for nitrogen cycling in soil due to their good symbiotic coexistence with rhizobia bacteria, which are well known for their nitrogen-fixing capacities. The use of legumes for phytoremediation of pesticides have been affirmed by many researchers such as Fan et al., (2008); Hamdi et al., (2012); Carrasco-Gil et al., (2013) and Li et al., (2013), it may therefore be more feasible to bio-augment such mechanism with adequate microorganisms that can also bio-remediate for faster and effective remediation.

Symbiotic association between plants and microorganisms can be exploited for faster soil clean technology and this has been previously employed for remediation of several contaminants. This has been affirmed to be effective and essential for ecological sustainability. The pH, nutrient, profile, texture, density and the physio-chemical attributes of the polluted soil couple with the bioavailability and toxicity of the targeted soil pollutant, diversity, trait/attribute and genetic make-up of particular plant and the native/introduced soil microbial species are key factors to be considered in setting up a feasible synergistic plant-microbial remediation mechanism (Segura et al., 2009). Having established these factors, this mechanism is feasible for sustainable pesticide remediation through rhizospheric actions of the two entities (Abhilash et al., 2012).

2. Synergistic Plant-Microbial Biodegradation of Pesticide Pollutants

Synergistic remediation involves the complex interactions between the plant roots, root exudates, rhizosphere soil and microbes for better biological responses and degradation of soil contaminants to nontoxic compounds. This has been reported to results in increased number, diversity and metabolic activity of plants and microbes which are able to degrade contaminants, it also support plant growth in close vicinity of roots compared to bulk soil (Gkorezis et al., 2016). It is also established that the rate at which pollutant degrade during plant-microbial interaction is higher than the sole contribution of single entity in remediation processes (Khan et al., 2013b). The interaction between plants and microbes having catabolic genes, has led to the evolution of a diverse variety of catabolic enzymes that can metabolize and detoxify the xenobiotics such as pesticides (Hong et al., 2015; Xun et al., 2015). The combined effect of plants and microbes has been reported to enhance the phytoremediation of soil and water contaminated with different organic pollutants (Khan et al., 2013b; Shehzadi et al., 2014). Likewise, several studies were performed to explore the potential of plant–microbe partnership for the remediation of POPs contaminated soil and water (Jha and Jha 2015). Examples of such microorganisms which take part in synergistic remediation are itemized below:

2.1. Plant Growth-Promoting Bacteria

Plant growth-promoting bacteria (PGPR) have the capability to act as biocontrol agents against plant pathogens, as well stimulate plant growth, which may be due to one of the mechanisms such as antibiotic production, depletion of iron from the rhizosphere, induced systemic resistance, production of fungal cell wall lysing enzymes, and competition for binding sites on the root (Glick et al., 2007). Attention has been shifted to bioremediation with PGPR despite their use in enhancing plant growth and as biocontrol agents against plant diseases (Huang et al., 2005). They degrade and/or mineralize organic compounds in association with plants, as against inorganic compounds. Most of these bacteria cannot survive in the near-starvation conditions found in soils, including the rhizosphere (Zhuang et al., 2007).

2.2. Endophytic Microorganisms

Endophytic microorganisms colonize the internal tissues of plants causing symptomless infections which can exist throughout the plant, these are usually ignored due to invisible damage to the host plant (Olowe et al., 2015, 2017). It has been reported that the route of colonization, through which the endophytes gain entry into plants is via the roots, with the exception of seed endophytes, (Germaine et al., 2004). They either reside in specific plant tissues like the root cortex or the xylem or colonize the plant systematically by transport through the vascular system or apoplast (Quadt-Hall mann et al., 1997). Endophytic bacteria and fungi have been isolated from a variety of healthy plant species such as different grass species and maize (Dalton et al., 2004; Olowe et al., 2015, 2017). Additionally to their beneficial effects on plant growth, endophytes have been use to enhance the efficiency of phytoremediation (Weyens et al., 2009).

2.3. Mycorrhiza

Arbuscular mycorrhizal fungi (AMF) establish symbiotic relationships with land plants and as a result, they are used in phytoremediation technology (Khan 2006). AMF modifies pollutant bioavailablity in several ways, this include: competition with roots and other microorganisms for water and pollutant uptake, protection of roots from direct interaction with the pollutant via formation of the ectomycorrhizal sheath, and impeded pollutant transport through increased soil hydrophobicity (Meharg and Cairney, 2000). Ectomycorrhizal associations has been reported to display considerable resistance against toxicity in soil (Meharg and Cairney, 2000). The fungal sheath, the density and surface area of the mycelium are probable characteristics to determine the impact of an ectomycorrhizal association on metal toxicity and protecting the host plant from pollutant contact (Hartley et al., 1997). The mycorrhizae contribute to the resistance of plant–microbial interactions via enhanced degradation of organic pollutants in the mycorrhizosphere which as result lower the bioavailable concentration of the pollutant in soil (Meharg and Cairney, 2000).

2.4. Roles of Bioaugmentation

Bioaugmentation is the deliberate inoculation of microorganisms, characterized with desired catalytic potentials to enhance the capacity of a contaminated environment (Feng et al., 2017). The selection of the appropriate microbial strains for bioaugmentation is dependent on a high potential for contaminant degradation, fast growth, ease of cultivation, the ability to withstand high concentrations of pollutants and to survive in a wide range of environmental conditions (Mariusz et al., 2016). Autochthonous, allochthonous and gene bioaugmentation are approaches distinguished based on the origin of inoculants (Figure 1). In autochthonous; microorganisms are isolated from the contaminated environments and re-injected onto the same site, allochthonous bioaugmentation; microorganisms are recruited from another site after which they are introduced onto the polluted site (Semrany et al., 2012; Mariusz et al., 2016), while thethe gene bioaugmentation involves the use of genetically engineered microorganisms (GEMs) equipped with genes encoding the enzymes responsible for some desired functions as well as the introduction of catabolic vectors directly into the environment (Gao et al., 2015; Mariusz et al. 2016). For microorganisms to be used effectively in bioaugmentation process, it is necessary to determine the microbial potential for the removal of pesticides under optimal conditions in liquid media. It has been confirmed that bacteria and fungi that belong to a wide range of genera and that are used in bioaugmentation are capable of degrading pesticides from different chemical classes in liquid cultures (Table 1). Despite its success, there is a plethora of barriers such as adaptations of microbial strains or microbial consortia, competition between introduced and indigenous biota, use of other organic substrates instead of the pollutant, and the availability of the contaminant to the microorganisms, determine the activity, persistence, and performance of bioaugmented inoculants (Adams et al., 2015). However, it is necessary that identification and isolation of strains should be made based on the comprehensive understanding of the composition, population dynamics, physiological profiles of the source communities, and those in the target habitats (Thompson et al., 2005).

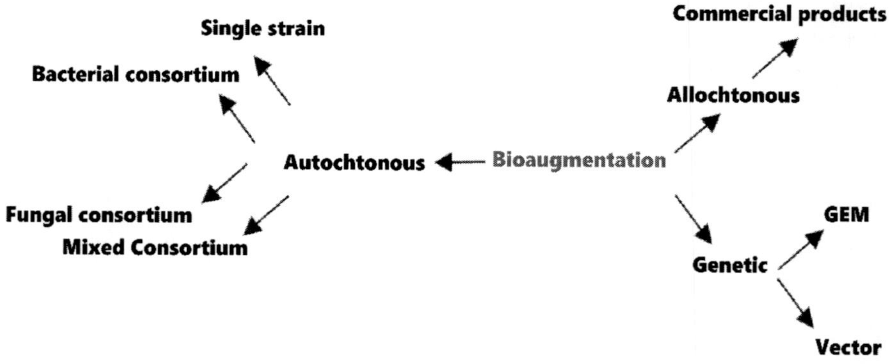

Figure 1. Types of bioaugmentation as one of the engineered bioremediation strategies.

Table 1. Studies on pesticide-degrading microorganisms for the bioaugmentation of contaminated sites

Pesticide	Microorganism
DDT	*Alcaligenes eutrophus, Aerobacter aerogenes*
	Sphingobacterium sp.
	Penicillium miczynskii, Aspergillus sydowii, Trichoderma sp., *Penicillium raistrickii, Aspergillus sydowii and Bionectria* sp.
	Aerobacter aerogenes
	Trichoderma viridae, Pseudomonas sp., *Micrococcus* sp., *Arthrobacter* sp., *Bacillus* sp.
	Pseudomonas sp.
	Sphingobacterium sp.
Dieldrin	*Pseudomonas* sp.
Endosulfan	*Pseudomonas aeruginosa, Burkholderia cepaeia, Arthrobacter sp.* KW, *Aspergillus niger*
PCP	*Arthrobacter* sp., *Flavobacterium* sp.
1,4- Dichlorobenzene	*Pseudomonas* sp.
Lindane	*Sphingomonas paucimobilis*
	Pleurotus ostreatus, Streptomyces sp.
	Ganodermaaustrale
DDE	*Phanerochaete chrysosporium*
DDD	*Trichoderma* sp
Heptachlor epoxide	*Phlebia* sp.
	Phanerochaete chrysosporium,
Heptachlor O	*Phlebia* sp.
	Bjerkandera sp.
	Trichoderma viridae, Pseudomonas sp., *Micrococcus* sp., *Bacillus* sp.
Toxaphene O	*Trichoderma viridae, Pseudomonas* sp., *Micrococcus* sp.
	Arthrobacter sp., Bacillus sp
Aldrin O	*Pseudomonas* sp. 105
	Pseudomonas sp.

Source: Parte et al., (2017).

2.5. Roles of Biostimulation

This is done by adding various growth-promoting nutrients such as carbon, nitrogen, and phosphorus to stimulate the plants and indigenous microbes degrading organic pollutants (Adams et al., 2015). In some cases, biostimulation can be achieved using selected supplemental nutrients such as crude plant extracts or plant exudates that act as co-substrates inducing enzymes for co-metabolism (Jha et al., 2015). Additives such as electron acceptors or donors, oxidants, and bulking agents can also be used in biostimulation (Kuppusamy et al., 2016). Alternatively, successful biostimulation can be obtained by generating an optimal balance of physical factors such as aeration, temperature, and buffering of environmental pH by altering the redox state and electrokinetic state of contaminated soils (Pandey et al., 2009). The challenges include the delivery of additives in a manner readily available to indigenous microbial communities, and the density and diversity of indigenous microbes capable of degrading high concentrations of pollutants (Adams et al., 2015). Biostimulation can be more effective when combined with bioaugmentation methods. Both indigenous and exogenous micro-organisms can benefit from biostimulation by the addition of energy sources or electron acceptors (El Fantroussi and Agathos, 2005). However, any such planned intervention must be followed by site-specific evaluations of the contaminated site.

2.6. Roles of Biosurfactants

The synthetic surfactant in pesticides industries act as emulsifying, dispersing, spreading and wetting agents, enhance the efficiency of pesticides and are used–as insecticides which gives them defensive properties (Rostas and Blassmann 2009). There are different types of surfactants such as anionic, cationic, amphoteric, and nonionic, these are used in several pesticide manufacturing industries (Mulqueen 2003). Pesticide residues are known to persist for years in soil, spreads in air and water, on surfaces of fruit and vegetables (Street 1969). The synthetic surfactants are reported as potent organic pollutants in soil (Petrovic and Barcelo 2004). Considering the adverse effect of pesticides and surfactants, associated with the pesticides, there is need to use the eco-friendly biosurfactants to replace the harmful surfactants in the pesticide industries thus preventing pollutants (Hopkinson et al., 1997). Research related to exploration of soil bacteria, which can utilize the chemical surfactant in agriculture soil as carbon source, can be another alternative to this environmental problem. There is a report on bacteria belonging to

Pseudomonas sp. and *Burkholderia* sp. from paddy field to degrade surfactants (Nishio et al., 2002). Agriculture important products like pesticides formed with the assistance of biosurfactant can be widely used on agricultural fields. The need for agrochemical industries is to develop effective formulation technology and to achieve this goal; many companies can employ mixture of biosurfactants in different combinations with the polymers to make excellent formulations for agricultural applications.

2.7. Roles of Enzymes

Enzymes usually play key roles in biodegradation of any xenobiotics, they catalysis/speed up pollutant degradation to a noticeable rate in other to restore polluted environment (Rao et al., 2010). In addition, they are involved in the degradation of pesticide compounds, both in the target organism (detoxification mechanisms and evolved metabolic resistance) and in the wider environment (biodegradation by soil and water microorganisms). *Pseudomonas putida* for example produce theoretical oxygen demand (TOD) enzyme as a representative of a much larger family of enzymes with novel application in the biocatalysis of environmentally relevant reactions. Fungal enzymes especially, oxidoreductases, laccase and peroxidases have prominent applications in removal of polyaromatic hydrocarbons (PAHs) contamination either in fresh, marine water or terrestrial (Balaji et al., 2014).

Table 2. Some of the selected sources of enzymes and their degradation property

Enzyme	Source	Degradation
Arly aclyamidase	*Bacillus sphaericus*	Herbicide and fungicide
Organophosphorus hydrolase (OPH)	*B.diminuta* and *Flavobacterium sp.*	Xenobiotics compounds
Organophosphorus acid anhydrolase (OPAA)	*Alteromonas undina* and *Alteromonas haloplanktis*	Xenobiotics compounds

Adapted from Uqab et al., (2016).

3. Genetic Inferences of Plant-Microbial Interactions during Bioremediation

3.1. Genetics for Pesticide Degradation

The role of catabolic genes and the application of recombinant DNA technology cannot be over-emphasized in the biodegradation of pesticides. Knowledge of genetic basis of the mechanisms of biodegradation of pollutants by microoganisms and interaction with the environment is important for successful implementation of the technology for *in situ* remediation (Hussain et al., 2009). This involves the isolation of pesticide degrading microorganisms and the characterization of genes encoding pesticide degradation enzymes which can yield unique insights into the molecular events that lead to the development of enhanced pesticide degradation phenomenon. Microbial enzymes with the capacity to hydrolyze pesticides such as organophosphorus hydrolase (OPH; encoded by the *opd* gene) have been reported (Yan et al., 2007). The *opd* gene has been found in bacterial strains that can utilize organophosphate pesticides as carbon source. This gene was reported to have 996 nucleotides, a typical promoter sequence of the promoter TTGCAA N17 TATACT from *E. coli* (Zhang et al., 2005).

3.2. High-Throughput Approaches to Characterize Contaminant Sites and Monitoring Bioremediation

3.2.1. Fingerprinting Techniques

Genetic fingerprinting techniques generate a specific pattern of a particular microbial community which are based on direct analysis of amplified PCR products from environmental DNA using universal or specific primers (Nocker et al., 2007). Some of these techniques include terminal restriction fragment length polymorphism (T-RFLP), length heterogeneity analysis by PCR (LH-PCR), single strand conformation polymorphism (SSCP), denaturing gradient gel electrophoresis (DGGE), or ribosomal intergenic spacer analysis (RISA) and produce a community fingerprint based on either sequence polymorphism or length polymorphism.

Table 4. Genes with the ability to degrade pesticides

Organism	Gene	Organism
Bacteria	*opd*	*Pseudomonas diminuta*
	opaA	*Alteromonas* spp.
	opdA	*A. radiobacter*
	adpB	*Nocardia* sp.
	pepA	*Escherichia coli*
	hocA	*Pseudomonas monteilli*
	pehA	*Burkhlderia caryophilli*
	phn	*Bacillus cereus*
	ophB	*Burkholderia* sp. JBA3.
	ophC2	*Stenotrophomonas* sp. SMSP-1.
	opdB	*Lactobacillus brevis*
	imh	*Arthrobacter* sp. scl-2
	mpd	*Ochrobactrum* sp. Yw28, *Rhizobium radiobacter*
	oph	*Arthobacter* sp
	mpdB	*Arthrobacter* sp.
	mpdB	*Burkholderia cepacia*
	opdE	*Enterobacter* sp.
Fungi	*A-opd*	*Aspergillus niger*
	P-opd	*Penicillium lilacinum*

(Adapted from Singh and Walker, 2006).

3.2.2. Quantitative/Real Time PCR

Quantitative PCR (Q-PCR), or real-time PCR, has been used in microbial investigations to measure the abundance and expression of taxonomic and functional gene (Bustin et al., 2005; Smith and Osborn 2009). Unlike traditional PCR, which relies on end-point detection of amplified genes, Q-PCR uses either intercalating fluorescent dyes such as SYBR Green or fluorescent probes (TaqMan) to measure the accumulation of amplicons in real time during each cycle of the PCR. When Q-PCR is coupled with a preceding reverse transcription (RT) reaction, it can be used to quantify gene expression (RT-Q-PCR). Several sets of 16S and 5.8S rRNA gene primers have been designed for rapid Q-PCR based quantification of soil bacterial and fungal microbial communities (Fierer et al., 2005). Q-PCR has also been successfully used in environmental samples for quantitative detection of important physiological groups of bacteria such as ammonia oxidizers, methane oxidizers, and sulfate reducers by targeting amoA, pmoA, and dsrA genes, respectively (Foti et al., 2007).

3.2.3. DNA Microarrays

DNA microarrays have been used primarily to provide a high-throughput and comprehensive view of microbial communities in environmental samples. The PCR products amplified from total environmental DNA is directly hybridized to known molecular probes, which are attached on the microarrays (Zhou, 2003; Gentry et al., 2006). Unlike the conventional nucleic acid membrane hybridization, microarrays present the advantage of miniaturization (thousands of probes can be spotted on a slide), high sensitivity and rapid ("real-time") detection (Eyers et al., 2004b). Hybridization signal intensity on microarrays is directly proportional to the abundance of the target organism. Cross hybridization is a major limitation of microarray technology, particularly when dealing with environmental samples. In addition, the microarray is not useful in identifying and detecting novel prokaryotic taxa. The ecological importance of a genus could be completely ignored if the genus does not have a corresponding probe on the microarray. DNA microarrays used in microbial ecology could be classified into two major categories depending on the probes, these are: 16S rRNA gene microarrays which contains about 30,000 probes of 16S rRNA gene targeted to several cultured microbial species and have been used for rapid profiling of environmental microbial communities during bioterrorism surveillance, bioremediation, climate change, and source tracking of pathogen contamination (Brodie et al., 2007). The functional gene arrays (FGA) are designed primarily to detect specific metabolic groups of bacteria and shed light on the *in situ* community metabolic potential.

3.2.4. Metagenomics

Metagenomics is the assessment of the microbial genomes of a given habitat, which does not rely on prior knowledge of the microbial habitats (Eyers et al., 2004). It refers to the construction of metagenomic libraries which are; generation of DNA fragments of appropriate size, ligation of the fragments into an appropriate cloning vector (e.g., cosmid, fosmid or bacterial artificial chromosome (BAC) vectors), introduction of the recombinant vectors into a suitable bacterial cloning host, and screening of clones harboring particular activities, or containing specific sequences (Daniel, 2005). However, the application of metagenomics to environmental samples is influenced with the development of various commercial kits to build the DNA libraries which is hampered with difficulties such as the isolation of genomic DNA from complex environmental matrices, such as soil, is particularly challenging in terms of quality, molecular size and adequate representation of all microbial genomes (Cowan et al., 2005).

3.2.5. Metatranscriptomics and Metaproteomics

These approaches help to understand the actively expressed genes in a microbial community and are good indicators for the microbial functions, which are expressed under conditions at the time of sampling (Stenuits et al., 2008). In metatranscriptomic, RNA is extracted from an environmental samples which is converted to cDNA and sequenced to metagenomic. This approach provides an inventory of the actively expressed genes in a sample (Stenuita et al., 2008). Metaproteomics deals with high-resolution mass spectrometry combined with enzymatic digests of proteins and liquid chromatography (Hettich et al., 2013). This gives insights into complement of proteins found in an environmental samples including post- translational modification that may affect their activity. Strengths and weaknesses of these approaches have been reviewed (Hettich et al., 2013; Xong et al., 2015).

4. Innovation and Developmental Impacts of Plant-Microbial Bioremediation Mechanism

Plant-microbial bioremediation has been established to positively enhanced soil remediation of different pollutants and this is currently gaining interests in the field of bioremediation. This method creates an environmental friendly and cost effective and it is more importantly feasible in a low scaled treatment (remediation of a small site space). This mechanism has been reported far back in early 1990s (Anderson et al., 1993; Brazil et al., 1995). Today, there have been several developments in the investigation of this innovative technology from bench-scale experiments towards addressing more in field remediation efficiencies. Fulekar (2014) reported enhanced rhizosphere bioremediation of pesticides (Chlorpyrifos, Cypermethrin and Fenvalerate) using some microbial consortium and plants (*Pennisetum pedicellatum* and *Cenchrus setigerus*) while Cherni et al., (2015) showed original insights into the effect of the herbicide on the rhizospheric area of *Medicago sativa* with or without rhizobial inoculation by the fact that glyphosate changes microbial diversity and affects soil enzymatic activities. Singh et al., (2016) established the synergistic influence of *Vetiveria zizanioides* and selected rhizospheric microbial strains on remediation of endosulfan contaminated soil.

References

Abhilash, P. C., Powell, J. R., Singh, H. B., and Singh, B. K., 2012. Plant–microbe interactions: novel applications for exploitation in multipurpose remediation technologies. *Trends Biotechnol.* 30: 416–420. doi: 10.1016/j.tibtech.2012. 04.004.

Adams, G.O., Fufeyin, P.T., Okoro, S.E., Ehinomen, I., 2015. Bioremediation, biostimulation and bioaugmention: a review. Int. *J. Environ. Biorem. Biodegrad.* 3: 28–39.

Alloway, B. J., and Trevors, J. T., 2013. *Heavy Metals in Soils-Trace Metals and Metalloids in Soils and their Bioavailability*. New York: Springer Dordrecht Heidelberg.

Anderson T.A., Guthrie E.A., Walton B.T., 1993. Bioremediation in the rhizosphere. *Environ Sci Technol.* 27: 2630–2636.

Asemoloye, M.D., Ahmad, R., Jonathan S.G., 2017a. Synergistic Rhizosphere Degradation of γ-Hexachlorocyclohexane (Lindane) through the Combinatorial Plant-Fungal Action. *Plos One*, 12(7) e-0183373.

Asemoloye M.D., Ahmad R., Jonathan S.G., 2017b. Synergistic Action of Rhizospheric Fungi with Megathyrsus maximus Root Speeds up Hydrocarbon Degradation Kinetics in Oil Polluted Soil. *Chemosphere*, 184: 1-12.

Asemoloye, M.D., Jonathan, S.G., Jayeola, A.A., Ahmad, R. 2017c. Mediational influence of spent mushroom compost on phyto-remediation of black-oil hydrocarbon polluted soil and response of *Megathyrsus maximus* Jacq. *Journal of Environmental Management* 200: 253-262.

Balaji, V., Arulazhagan, P., Ebenezer, P., 2014. Enzymatic bioremediation of polyaromatic hydrocarbons by fungal consortia enriched from petroleum contaminated soil and oil seeds. *J Environ Biol* 35: 521-529.

Brazil, G.M., Kenefick, L., Callanan, M., Haro, A., De Lorenzo, V., Dowling, D.N., Ogara, F., 1995. Construction of a rhizosphere Pseudomonad with potential to degrade polychlorinated-biphenyls and detection of BPH gene expression in the rhizosphere. *Appl Environ Microbiol.* 61: 1946–1952.

Brodie, E.L., DeSantis, T.Z., Joyner, D.C., Baek, S.M., Larsen, J.T., Andersen, G.L., et al., 2006. Application of a high-density oligonucleotide microarray approach to study bacterial population dynamics during uranium reduction and reoxidation. *Appl Environ Microbiol* 72(9): 6288–98.

Carrasco-Gil, S., Siebner, H., LeDuc, D. L., Webb, S. M., Millaìn, R., Andrews, J. C., et al., 2013. Mercury localization and speciation in plants grown hydroponically or in a natural environment. *Environ. Sci. Technol.* 47: 3082–3090. doi: 10.1021/es303310t.

Cherni, A. E., Trabelsi, D., Chebil, S., Barhoumi, F., Rodríguez-Llorente, I.D., Zribi, K., 2015. Effect of Glyphosate on Enzymatic Activities, Rhizobiaceae and Total Bacterial Communities in an Agricultural Tunisian Soil. *Water Air Soil Pollut* 226: 145.

Cowan, D., Meyer, Q., Stafford, W., Muyanga, S., Cameron, R., Wittwer, P., 2005. Metagenomic gene discovery: past, present and future. *Trends Biotechnol* 23: 321–9.

Dalton, D., Kramer, S., Azios, N., Fusaro, S., Cahill, E., Kennedy, C., 2004. Endophytic nitrogen fixation in dune grasses (Ammophila arenaria and Elymus mollis) from Oregon. *FEMS Microbiol Ecol.* 49:469-479.

Daniel, R., 2005. The metagenomics of soil. *Nat Rev Microbiol* 3: 470–8.

El Fantroussi, S., Agathos, S.N., 2005. Is bioaugmentation a feasible strategy for pollutant removal and site remediation? *Curr. Opin. Microbiol.* 8: 268–275.

Eyers, L., George, I., Schuler, L., Stenuit, B., Agathos, S.N., El Fantroussi, S., 2004. Environmental genomics: exploring the unmined richness of microbes to degrade xenobiotics. *Appl Microbiol Biotechnol* 66: 123–30.

Fan, S. X., Li, P. J., Gong, Z. Q., Ren, W. X., and He, N., 2008. Promotion of pyrene degradation in rhizosphere of alfalfa (*Medicago sativa* L.). *Chemosphere* 71, 1593–1598. doi:10.1016/j. Chemosphere.2007. 10.068.

Feng, N.-X., et al., 2017. Efficient phytoremediation of organic contaminants in soils using plant–endophyte partnerships, *Sci Total Environ*, http://dx.doi.org/10.1016/j.scitotenv.2017.01.07.

Foti, M., Sorokin, D.Y., Lomans, B., Mussman, M., Zacharova, E.E., Pimenov, N.V., Kuenen, J.G., Muyzer, G. 2007. Diversity, activity, and abundance of sulfate-reducing bacteria in saline and hypersaline soda lakes. *Appl. Environ. Microbiol.* 73: 2093–3000.

Fulekar, M.H., 2014. Rhizosphere Bioremediation of Pesticides by microbial consortium and potential microorganism. *International Journal of Current Microbiology and Applied Sciences.* 3(7): 235-248.

Gao, C., Jin, X., Ren, J., Fang, H., Yu, Y., 2015. Bioaugmentation of DDT-contaminated soil by dissemination of the catabolic plasmid pDOD. *J. Environ. Sci.* 27: 42–50.

Gentry, T.J., Wickham, G.S., Schadt, C.W., He, Z., Zhou, J., 2006. Microarray applications in microbial ecology research. *Microb Ecol* 52:159–75.

Germaine, K., Keogh, E., Borremans, B., van der Lelie, D., Barac, T., Oeyen, L., Vangronsveld, J., Porteus, Moore, F., Moore, E.R.B., Campbel, C.D., 2004. Colonization of poplar trees by gfp expressing endophytes. *FEMS Microbiol Ecol.* 48: 109-118.

Gkorezis, P. Daghio, M. Franzetti, A. Van Hamme J. D. Sillen, W. and Vangronsveld, J., 2016. The Interaction between Plants and Bacteria in the Remediation of Petroleum Hydrocarbons: An Environmental Perspective. *Front Microbiol.* 7: 183.

Glick, B.R., Todorovic, B., Czarny, J., Cheng, Z., Duan, J., McConkey, B., 2007. Promotion of plant growth by bacterial ACC deaminase. *Crit Rev Plant Sci.* 26: 227–42.

Hamdi, H., Benzarti, S., Aoyama, I., and Jedidi, N., 2012. Rehabilitation of degraded soils containing aged PAHs based on phytoremediation with alfalfa (Medicago sativa L.). *Int. Biodeter. Biodegr.* 67, 40–47. doi: 10.1016/j.ibiod.2011.10.009.

Hartley, J., Cairney, J.W.G., Meharg, A.A., 1997. Do ectomycorrhizal fungi exhibit adaptive tolerance to potentially toxic metals in the environment? *Plant Soil.* 189: 303–319.

Hettich, R.L., Pan, C.L., Chourey, K., Giannone, R.J., 2013. Meta-proteomic: harnessing the power of height performance mass spectrometry to identify the suite of proteins that control metabolic activities in microbial communities. *Anal Chem* 85: 4203- 4214. doi 10.1021/ac303053e.

Hong, Y., Liao, D., Chen, J., Khan, S., Su, J., Li, H., 2015. A comprehensive study of the impact of polycyclic aromatic hydrocarbons (PAHs) contamination on salt marsh plants *Spartina alterniflora*: implication for plant-microbe interactions in phytoremediation. *Environ Sci Pollut Res* 22: 7071–7081.

Hopkinson, M.J., Collins, H.M., Goss, G.R., 1997. Pesticide formulations and application systems: ASTM Committee E-35 on Pesticides. *ASTM International.* 17(1328):1–331.

Huang, X.D., El-Alawi, Y., Penrose, D.M., Glick, B.R., Greenberg, B.M., 2004. A multi-process phytoremediation system for removal of polycyclic aromatic hydrocarbons from contaminated soils. *Environ Pollut.* 130: 465–76.

Hussain, S., Siddique, T., Arshad, M., Saleem, M., 2009. Bioremediation and Phytoremediation of Pesticides: Recent Advances. *Critical Reviews in Environmental Science and Technology* 39(10): 843-907.

Jha, P., Jha, P., 2015. Plant-microbe partnerships for enhanced biodegradation of polychlorinated piphenyls. In: Arora NK (ed) *Plant microbes symbiosis. Applied facets*. Springer, India, pp 95–110.

Khan, A.G., 2006. Mycorrhizoremediation - an enhanced form of phytoremediation. *J Zhejiang Univ Sci B.*7:503–14.

Khan, S., Afzal, M., Iqbal, S., Khan, Q.M., 2013a. Plant–bacteria partnerships for the remediation of hydrocarbon contaminated soils. *Chemosphere* 90: 1317–1332.

Khan, S., Afzal, M., Iqbal, S., Mirza, M.S., Khan, Q.M., 2013b. Inoculum pretreatment affects bacterial survival, activity and catabolic gene expression during phytoremediation of diesel contaminated soil. *Chemosphere* 91: 663–668.

Li, Y., Liang, F., Zhu, Y. F., and Wang, F. P., 2013. Phytoremediation of a PCB contaminated soil by alfalfa and tall fescue single and mixed plants cultivation. *J. Soil. Sediment.* 13, 925–931. doi: 10.1007/s11368-012-0618-6.

Lohmann, R., Breivik, K., Dachs, J., and Muir, D., 2007. Global fate of POPs: current and future research directions. *Environ. Pollut.* 150, 150–165. doi: 10.1016/j.envpol.2007.06.051.

Meharg, A.A., Cairney, J.W.G., 2000. Extomycorrhizas- extending the capabilities of rhizosphere remediation. *Soil Biol Biochem.* 32:1475–1484.

Mulqueen, P., 2003. Recent advances in agrochemical formulations. *Adv Colloid Interface Sci.* 106: 83–107.

Nishio, E., Ichiki, Y., Tamura, H., Morita, S., Watanabe, K., Yoshikawa, H., 2002. Isolation of bacterial strains that produce the endocrine disruptor, octylphenol diethoxylates, in paddy fields. *Biosci Biotechnol Biochem.* 66:1792–1798.

Nocker, A., Burn, M., Camper, A.K., 2007. Genotypic microbial community profiling: A critical technical review. *Microb Ecol* 54: 276–89.

Olowe, O.M. Odebode, A.C. Olawuyi, O.J. Sobowale, A.A., 2017. Molecular Variability of *Fusarium verticillioides* (Sacc.) in Maize from Three Agro-Ecological Zones of Southwest Nigeria. *American Journal of Molecular Biology*, 7: 30-40

Olowe, O.M., Olawuyi, O.J., Odebode, A.C., 2015. Response of Maize Genotypes to *Fusarium verticillioides* Strains from Two Agro Ecological Zones in Southwest Nigeria. *International Journal of Pure and Applied Sciences and Technology*, 27: 77-86.

Pandey, J., Chauhan, A., Jain, R.K., 2009. Integrative approaches for assessing the ecological sustainability of in situ bioremediation. *FEMS Microbiol.* Rev. 33: 324–375.

Parte, S. G., Mohekar, A. D and Kharat, A. S. (2017). Microbial degradation of pesticide: A review. *African Journal of Microbiology research.* 11(24): 992-1012.

Petrovic, M., Barcelo, D., 2004. Analysis and fate of surfactants in sludge and sludge-amended soil. *Trends Analyt Chem.* 23:10–11.

Petrovic, M., Barcelo, D., 2004. Analysis and fate of surfactants in sludge and sludge-amended soil. *Trends Analyt Chem.* 23: 10–11.

Quadt-Hallmann, A., Benhamou, N., Kloepper, J.W., 1997. Bacterial endophytes in cotton: mechanisms entering the plant. *Can J Microbiol.* 43:577-582.

Rajkumar, M., Sandhya, S., Prasad, M. N. V., and Freitas, H., 2012. Perspectives of plant-associated microbes in heavy metal phytoremediation. *Biotechnol. Adv.* 30, 1562–1574. doi: 10.1016/j.bio techadv.2012.04.011.

Rao, M.A., Scelza, S.R., Gianfreda, L., 2010. Role of enzymes in the remediation of polluted environments. *J Soil Sci Plant Nutr* 10: 333-353.

Rostas, M., Blassmann, K., 2009. Insects had it first: surfactants as a defense against predators. *Proc R Soc B.* 276:633–638.

Satish, G.. Parte, Ashokrao, D., Mohekar, A., Kharat S., 2017. Microbial degradation of pesticide: A review. *Afr. J. Microbiol. Res*, 11(24): 992-1012.

Segura, A., and Ramos, J. L., 2013. Plant–bacteria interactions in the removal of pollutants. *Curr. Opin. Biotech.* 24, 467–473. doi: 10.1016/j.copbio.2012.09.011

Segura, A., Rodríguez-Conde, S., Ramos, C., and Ramos, J. L., 2009. Bacterial responses and interactions with plants during rhizoremediation. *Microb. Biotechnol.* 2, 452–464. doi: 10.1111/j.1751 -7915.2009.00113.x.

Semrany, S., Favier, L., Djelal, H., Taha, S., Amrane, A., 2012. Bioaugmentation: possible solution in the 879 treatment of bio-refractory organic compounds. *Biochem. Eng. J.* 69: 75–86.

Shehzadi, M., Afzal, M., Khan, M.U., Islam, E., Mobin, A., Anwar, S., Khan, Q.M., 2014. Enhanced degradation of textile effluent in constructed wetland system using Typha domingensis and textile effluent-degrading endophytic bacteria. *Water Res* 58: 152–159.

Singh, B.K., Walker, A., 2006. Microbial degradation of organo-phosphorus compounds. *FEMS Microbiol. Rev.* 30 (3): 428–471.

Singh, V., Singh, P., Singh, N., 2016. Synergistic influence of *Vetiveria zizanioides* and selected rhizospheric microbial strains on remediation of endosulfan contaminated soil. *Ecotoxicology*. 25: 1327-1337.

Smith, C.J., Osborn, A.M., 2009. Advantages and limitations of quantitative PCR (QPCR)-based approaches in microbial ecology. *FEMS Microbiol. Ecol.* 67 (1): 6– 20.

Stenuit, B. Eyers, L. Schuler, L. Agathos, S. N. George, I., 2008. Emerging high-throughput approaches to analyze bioremediation of sites contaminated with hazardous and/or recalcitrant wastes. *Biotechnology Advances* 26: 561–575.

Street, J.C., 1969. Methods of removal of pesticides residues. *Canad Med Ass J.* 100:154–160.

Teng, Y., Xu, Z. H., Luo, Y. M., and Reverchon, F., 2012. How do persistent organic pollutants be coupled with biogeochemical cycles of carbon and nutrients in terrestrial ecosystems under global climate change? *J. Soil. Sediment.* 12: 411–419. doi: 10.1007/s11368-011-0462-0.

Thompson, I.P., van der Gast, C.J., Ciric, L., Singer, A.C., 2005. Bioaugmentation for bioremediation: the challenge of strain selection. *Environ. Microbiol.* 7, 909–915.

Uqab, B., Mudasir, S., Nazir, R., 2016. Review on Bioremediation of Pesticides. *J Bioremed Biodeg* 7: 343. doi:10.4172/2155-6199. 1000343.

Valentín, L., Nousiainen, A., and Mikkonen, A., 2013. "Introduction to organic contaminants in soil: concepts and risks," in *Emerging Organic Contaminants in Sludges: The Handbook of Environmental Chemistry,* eds A. G. Kostianoy and D. Barceló (Berlin, Heidelberg: Springer-Verlag), 1:29.

Weyens, N, van de, D., Lelie, S., Taghavi, J., Vangronsveld. J., 2009. Phytoremediation: plant–endophyte partnerships take the challenge. *Curr Opin Biotechnol*. 20:1–7.

Xiong, W.L., Abraham, P.E., Li, Z., Pan, C.L., Hettich, R.L., 2015. Microbial metaproteomics for characterizing the range of metabolic functions and activities of human gut microbiota. *Proteomic* 15: 3424-3438. doi:10.1002/PMIC.2014400571.

Xun, F., Xie, B., Liu, S., Guo, C., 2015. Effect of plant growth-promoting bacteria (PGPR) and arbuscular mycorrhizal fungi (AMF) inoculation on oats in saline-alkali soil contaminated by petroleum to enhance phytoremediation. *Environ Sci Pollut Res* 22: 598–608.

Yan, Q.X., Hong, Q., Han, P., Dong, X.J., Shen, Y.J., Li, S.P., 2007. Isolation and characterization of a carbofuran-degrading strain *Novosphingobium* sp. FND-3. *FEMS Microbiol Lett.* 2007; 271: 207-213.

Zhang, R., Cui, Z., Jiang, J., Gu, X., Li, S., 2005. Diversity of organophosphorus pesticides degrading bacteria in a polluted soil and conservation of their organophosphorus hydrolase genes. *Can. J. Microbiol.* 5: 337-343.

Zhou, J., 2003. Microarrays for bacterial detection and microbial community analysis. *Curr Opin Microbiol* 6:288–94.

Zhuang, X., Chen, J., Shim, H., Bai, Z., 2007. New advances in plant growth-promoting rhizobacteria for bioremediation. *Environ. Int.* 33: 406–413.

Biographical Sketch

Michael Dare Asemoloye

Affiliation: Department of Botany, University of Ibadan

Education: BSc, MSc, Botany.

Research and Professional Experience:

Mr. ASEMOLOYE, Michael Dare is a researcher and a Doctoral Student (Ph.D.) in the Mycology/Biotechnology Department of Botany, University of Ibadan, Nigeria under a Postgraduate Scholarship, designating him as the University Postgraduate Scholar. He obtained his B.Sc., M.Sc. from the same Department. His major area of research interest is majorly on Food & Environmental Mycology/Biotechnology with interests in Rhizosphere clean-technology, Synergistic bio-remediation, Fungi proteomics, Metagenomic/metatranscriptomics, SMC

Technology, Mycoenzymology, Mycobiocontrol and so on. He has publications in highly referred national and international journals.

Professional Appointments:

He has won several notable awards/achievements including the 2015 The World Academy of Science (TWAS) and COMSATS Institute of Information Technology, Abbottabad Scholarship, the University of Ibadan Postgraduate Scholarship 2015, 2016 and 2017, University TA Scheme 2014 and 2015, UCLM Award 2016, and OGCC Young Scientist Award etc.

Publications from the Last 3 Years:

1. Jonathan S.G., Asemoloye MD.* and Ahmad R. 2017. *Synergistic Plant-Microbial Rhizosphere Mechanism: A Potential Headway for Remediation of Hydrocarbon Polluted Soils.* Available at SSRN: https://ssrn.com/abstract=3086229.
2. Asemoloye M.D*, Olowe O.M., Zaffar H., Oshibanj O.O. and Jonathan S.G. 2017. *Organic Compost as Catalyst or Mediator for Speedy and Cost Effective Bioremediation*. SSRN in Press. Available Online http://ssrn.com/abstract=3071988.
3. Asemoloye M.D., Ahmad R. and Jonathan S.G. 2017. Transcriptomic Responses of Catalase, Peroxidase and Laccase Encoding Genes and Enzymatic Activities of Oil Spill Inhabiting Rhizospheric Fungal Strains. *Environmental Pollution*, 230: 150-160. Asemoloye M.D., Ahmad R. and Jonathan S.G. 2017. Synergistic Rhizosphere Degradation of γ-Hexachlorocyclohexane (Lindane) through the Combinatorial Plant-Fungal Action. *Plos One*, 12(7) e-0183373.
4. Asemoloye M.D*, Ahmad R. and Jonathan S.G. 2017. Synergistic Action of Rhizospheric Fungi with Megathyrsus maximus Root Speeds up Hydrocarbon Degradation Kinetics in Oil Polluted Soil. *Chemosphere*, 184: 1-12.

5. Asemoloye M.D*, Jonathan S.G., Jayeola A.A., Ahmad R. 2017. Mediational influence of spent mushroom compost on phytoremediation of black-oil hydrocarbon polluted soil and response of *Megathyrsus maximus* Jacq. *Journal of Environmental Management* 200: 253-262.
6. Jonathan S.G, Bello T.S. and Asemoloye M.D*. 2017. Food Values, Spoilage Moulds and Aflatoxin Detection in 'Attiéké' (A Cassava Fermented Product). *J Microb Biochem Technol* 9:244-248. doi: 10.4172/1948-5948.1000372.
7. Asemoloye M.D., Olowe O.M. and Jonathan S.G. 2018 Clean-Biotechnology with Enhanced Plant-Microbial Synergistic Action: a Novel and Functional Remediation Mechanism for Pesticide Polluted Soils. In- *Soil Contamination: Sources, Assessment and Remediation*. Nova Science Publishers, Inc. NY, USA

In: Soil Contamination
Editor: Jesper E. Lund
ISBN: 978-1-53613-266-3

Chapter 6

THE IMPROVEMENT OF FOOD AND SOIL QUALITY USING BIOCHAR OR ZEOLITE

***Z. Ioannou*[1,*] *and V. Kavvadias*[2]**

[1]Hellenic Agricultural Organization "Demeter," Institute of Industrial and Forage Crops, Larissa, Greece

[2]Hellenic Agricultural Organization "Demeter," Institute of Soil and Water Resources, Department of Soil Science of Athens, Lycovrissi, Attiki, Greece

ABSTRACT

Biochar is a form of char and charcoal made from the partial combustion of organic materials. Its properties include high water-holding capacity, large surface area, cation exchange capacity, impact especially on microbial communities. These properties are used in agriculture, composting, and land remediation/restoration. Zeolites, natural and synthetic, are microporous aluminosilicate minerals that contain a wide variety of cations, such as Na^+, K^+, Ca^{2+}, Mg^{2+} and others.

* Corresponding Author Email:zioan@teemail.gr.

These ions can easily be exchanged for others in a solution. They have been widely studied regarding their suitability to be used in many different environmental applications worldwide, especially in agriculture. Some of the characteristics of zeolites that make them potentially desirable for improving the properties of soils are a large internal porosity that results in water retention, a uniform particle-size distribution that allows them to be easily incorporated, and a high cation-exchange capacity that enables them to retain nutrients. In literature, there are many published research works relative to the application of biochar to the remediation of soils contaminated with both organic and inorganic contaminants. Heavy metals can be stabilized in soil by adding biochar. It also contains high levels of organic carbon and enriches soils by adding nutrients such as potassium and calcium. Moreover, it increases soil pH, helping neutralize acidic soils, increasing plant and microbial growth, nutrient retention, water holding capacity, preventing groundwater eutrophication and reducing fertilizers requirements. It functions as a filter for the removal of traces of pharmaceutical drug residues and hormones from treated wastewater. Zeolite can be used as a liner for landfills. The ability of zeolites to immobilize contaminants in polluted soils has been also extensively tested. Reducing the plant availability of heavy metals (Cd, Pb, Cr, Zn, Cu, etc.) in soils is critical for optimizing agricultural production in areas with heavy metal contaminated soils. Phytoavailability of heavy metals correlates best with their concentrations in soil solution rather than with their total content in soil. The removal of heavy metals in polluted areas is very difficult because they persist in soils for very long periods. However, the fixation of heavy metals in a nonavailable form could be a useful method for soils that are already contaminated by heavy metals. Heavy metal phytoavailability may be reduced if the metals are sorbed or precipitated from the soil solution. One of the ways to heavy metal immobilization may be the application of zeolites. Zeolites in general have large cation-exchange capacity and expectedly attract positive-charged ions and, therefore, are widely used for sequestration of cationic pollutants like heavy metals. Natural and artificial zeolites increase ion-exchange sites in soils in addition to offering absorption sites for small molecules, due to their porous structure. Zeolites are able to retain heavy metals in soil. On the other hand, despite promising results for immobilization of contaminants by zeolites, several studies have demonstrated that zeolites addition failed to promote satisfactory growth of plants grown in contaminated soils. Thus, of serious concern within contaminated land remediation should be the possibility of 'Na toxicity' to the surrounding soil, and hence uptake by plants and grazing animals. Consequently, the addition of biochar or zeolite improves crop yields and food quality increasing soil nutrients, adsorbing organic compounds and stabilizing heavy metals in soil.

Keywords: biochar; zeolite; heavy metals; organic compounds; microbial activity

INTRODUCTION

The term of "biochar" derived from the soils of Amazon region, which are abundant in char-wood. These soils are well known as "Terra- Preta" or "Black - Earth" soils. "Terra-Preta" soils present high fertility, with increased amounts of nutrients such as nitrogen, phosphorus, potassium, calcium and soil organic matter due to the presence of black carbon (Glaser et al., 2001). Biochar is a solid material, i.e., a form of char and charcoal, made from the thermochemical conversion of organic materials in an oxygen-limited environment (International Biochar Initiative guidelines, 2013). The conversion process includes pyrolysis, gasification, torrefaction and hydrothermal carbonization. Pyrolysis includes the heat of the organic material (e.g., biomass) in the absence of oxygen conditions while gasification includes the heat of the organic material in higher temperatures than pyrolysis but in limited oxygen conditions. Torrefaction is a process where biomass is heated in an inert atmosphere at temperatures around 200-300 °C, while hydrothermal carbonization is a heating process at lower temperatures than torrefaction but in high pressure (2-6 MPa) (Kambo & Dutta, 2015; International Biochar Initiative, 2017). A new method that alternatively can be used for biochar production is microwave pyrolysis which offers rapid, uniform and selective heating of biochar compared to the previous processes (Wahi et al., 2017). The raw organic materials vary according to the physical and chemical properties and the final particle size of biochar. Different types of agricultural residues, i.e., algal biomass, forest residues, manures, activated sludge, energy crops and polymers, can be used as organic precursors (Oliveira et al., 2017; Simitzis and Ioannou, 2012). Number of factors influence biochars' production such as heating temperature and time, type of gas and probably the use of catalyst (Wahi et al., 2017). Biochar appears to be a new potential low-cost and effective material that has a wide range of characteristics transfusing unique properties. Its properties include high water-holding capacity, large specific surface area, enriched surface functional groups, cation exchange capacity, porous structure similar to that of activated carbons, impact especially on microbial communities (Tan et al., 2015). The above properties are in correlation with the biochars' production factors and are used in agriculture, composting, and land remediation/restoration.

The biochar application in soil achieves not only soil remediation from organic and inorganic pollutants but also ameliorates soil properties (Oliveira et al., 2017). It has several benefits including carbon sequestration to soil, enhancing soil fertility by improving nutrients and moisture retention and microbial activity increasing consequently crop productivity (Tang et al., 2013; Lone et al., 2015).

Zeolites are naturally occurring crystalline, hydrated aluminosilicates of alkali and alkaline earth cations (Gadepalle et al., 2007). They have a silicate framework consisting of interlocking tetrahedra of SiO_4 and AlO_4. The Si/Al ratio is inversely proportional to the cation content, however directly proportional to the thermal stability. Zeolites are one of the greatest cationic interchangers and their cationic interchange capacity is up to three times greater than other types of soil minerals (Ming and Boettinger, 2001). Cations can be exchanged by ion exchange and water can be removed reversibly by application of heat. The large vacant space in the zeolites allows space for large cation groups such as sodium, potassium, and even relatively large molecules and cation groups (ammonia, carbonate and nitrate ions) to enter as well as results in water retention. These channels allow the movement of the ions and molecules into and out of the structure. Natural zeolites are low-cost materials and are abundant in European territory and mainly in Mediterranean region (Inglezakis and Zorpas, 2012). The most well-known natural zeolites are clinoptilolite, erionite, chabazite, heulandite, mordenite, stilbite, and phillipsite (Kulasekaran and Damodar Reddy, 2011).They have good mechanical and thermal properties, high sorption capacity and they could adjust the pH of the soil or the aqueous solution. In addition they have a uniform particle-size distribution that allows them to be easily incorporated.

APPLICATIONS

Soil Remediation from Heavy Metals Using Biochar

The influence of biochar on the mobility and bioavailability of heavy metals has been investigated in field and laboratory experiments. Generally, inorganic metals (e.g., Mn^{2+}, As^{2+}, Pb^{2+},Cu^{2+},Ni^{2+},Cd^{2+},Hg^{2+}) are non-biodegradable and are toxic to fauna and flora at high concentrations (Lu et al., 2017). Biochar porous structure and surface functional groups interact with heavy metals in several ways such as ion exchange, complexation and adsorption (Oliveira et al., 2017). Researchers have shown that a high efficiency and low cost biochar supported nano-hydroxyapatite materials was used in the remediation of Pb-contaminated soil through batch (Yang et al., 2016a) and column experiments (Yang et al., 2016b). According to the results, the application of biochar enhances Pb immobilization, decreases the accumulation of Pb in plants and avoids the excessive release of phosphorus leading to the improvement of soil fertility (Yang et al., 2016a). After the addition of biochar, Pb concentration in the edible parts of cabbage mustard was less than the national vegetable tolerance requirement (Yang et al., 2016b). Another study (Su et al., 2016) has shown the use of biochar supported zero valent iron nanoparticles for the removal of Cr(VI) from contaminated soil, which was sown with cabbage mustard seeds. The immobilization process of hexavalent and total chromium succeeded totally. Macroscopically, such remediation ameliorates plant growth especially the length of root and stem decreasing simultaneously the phytotoxicity of Cr(VI) and Fe. Researchers approved the remediation of cadmium in soil by biochar –supported iron phosphate nanoparticles (Qiao et al., 2017) or vermicompost and biochar (Wang et al, 2017) improving soil fertility. According to Qiao et al., the produced composite effectively immobilize Cd in soil inhibiting Cd uptake to the belowground and aboveground parts of cabbage mustard increasing cabbage mustard growth. Wang et al., referred that the use of bio-conditioners, i.e., vermicompost and biochar, improves soil fertility and reduces bio-availability of Cd. A new study has shown the effectiveness of biochar to Cd-contaminated soil placed to the Campine region in Belgium and The Netherlands (Poucke et al., 2017). A reduction of Cd concentrations in soil solution appeared when biochar treatment was used in comparison with compost and peat treatments due to specific applied factors such as pH and metal complexation capacity of biochar. According to Vithanage et al., (2017), biochar provides changes in soil properties that may influence the bioavailability and phytotoxicity of As. Biochars, which derived from low temperature pyrolysis, are more effective to As adsorption. The adsorption mechanisms may include electrostatic attraction, ion-exchange, physical adsorption and chemical bonding. Modification in the production of biochar using activation processes, i.e activation by nitrogen or steam or acid, may facilitate As remediation (Vithanage et al., 2017). Biochar, which derived from chicken manure and oat hull, increased the immobilization of Cu in contaminated soils and enhanced plant growth (Moore et al., 2017). Another study indicated the reduction of the mobile form of Cu in soils with biochar and biochar-compost where plant species, i.e mustard, cress, ryegrass, were grown (Cárdenas-Aguiar, 2017). Wood biochar, wood ash and humic substances alone or in mixtures were applied on a heavily multi-contaminated sandy loam soil (Pukalchik et al., 2017). Wood ash reduced Cd and Zn mobility in soil while biochar was more suitable for Cu remediation to soils. Humic substances, which were mixed with wood ash or biochar, increased the immobilization efficiency (Cu, Cd, Zn) of amendments compared to wood ash or biochar alone (Pukalchik et al., 2017). Shen et al., examined the stability of Cu, Cd and Pb in soil washing residue, which contained biochar, under accelerated ageing (Shen et al., 2018). The addition of biochar to soil reduced the leached Cu, Cd and Pb from soil washing residue. A two step remediation strategy including soil washing and sludge derived biochar stabilization was applied for the reduction of Pb and Cu from contaminated soils (Yoo et al., 2017). The washing process extracted Cu and Pb from the two soils while the stabilization process decreased metal mobility and enhanced soil quality. The influence of biochars on the availability of heavy metals, e.g., Cd, Cu, Mn, Ni, Pb, Zn to ryegrass was examined (Zhang et al., 2016). According to the results, biochars decreased the uptake of Cd, Cu, Ni, Pb, Zn from plants except for Mn indicating a reduction of phytoavailability of heavy metals. Another study has shown that the application of rice straw biochar to a sandy loam soil contaminated with Cd, Cu, Pb, Zn immobilized heavy metals reducing their mobility and bioavailability (Lu et al., 2017). Techniques, that remediate the contaminated soils from heavy metals, can be divided to physical, chemical and biological techniques. The first one covers soil replacement, soil isolation, vitrification and electrokinetic remediation, the second one covers immobilization and soil washing while the third category covers phytovolatilization, phytostabilization, phytoextraction, chelate assisted phytoextraction and microbial assisted phytoextraction (Khalid et al., 2017).

Soil Remediation from Organic Compounds Using Biochar

Biochar has become a common tool for the immobilization of organic compounds in soil and the improvement of soil quality, crop yields and food quality. More specifically, wheat straw biochar, lignocellulosic substrate and mixtures of them with surfactant Brij30 were investigated as amendments for the remediation of polycyclic aromatic hydrocarbons (PAHs) from soil (Cao et al., 2016). According to the results wheat straw biochars inhibited the removal of phenanthrene and promoted the removal of benzo(a)pyrene in the polluted soils. Wheat straw biochars influenced negatively the activity of dehydrogenase and positively the activity of polyphenol oxidase. Researchers have shown that the combined use of biochar and compost led to the immobilization of polycyclic aromatic hydrocarbons (PAHs) and NSO-substituted PAH contaminants and facilitated the degradation of the contaminant fractions that were left (Sigmund et al., 2018). Another study examined the influence of biochar in 2,4-dichlorophenol and phenanthrene after 30 days in soil with/without the presence of the earthworm *Metaphire guillelmi* (Gu et al., 2016). Biochar application decreased the degradation and mineralization of the above pollutants and increased their metabolites in soil. Liu et al., (2017) investigated the bioremediation of cypermethrin-contaminated soil with the application of biochar and cypermethrin-degrading bacterial consortium in pot experiments. Biochar was used as a carrier for bacteria. The decrease in cypermethrin removal rate in soil and its bioavailability to bacteria was the result of the increased biochar addition. Koltowski et al., (2016) examined the application of biochars derived from willow, coconut and wheat straw, which was either activated or not, in three different soils with varying content of PAHs. Biochar activation reduced the bioavailable and bioaccessible PAHs content in comparison with non activated biochars. Biochars derived from straw were also used for the removal of dyes such as reactive brilliant blue (KNR) and rhodamine B (RB) indicating that biochars are effective adsorbents for both cationic and anionic dyes (Qiu et al., 2009). Rao et al., (2017) examined the use of biochars derived from poplar and conifer as remediation method of water and soil contamination by phenantrene and pentachlorophenol. The addition of biochar reduced phenantrene extractability and only higher amounts of it reduced significantly pentachlorophenol extractability. Different types of treated biochars derived from willow tree (*Salix viminalis*) and activated with microwave, carbon dioxide or superheated steam and untreared biochars were applied to three different industrially contaminated soils. Activated biochars increased the effectiveness of PAHs immobilization compared to untreated biochars (Koltowski et al., 2017). The remediation technologies that were used for the removal of heavy metals are similar to that of the removal of organic compounds, i.e chemical, physical and biological techniques (Morillo and Villaverde, 2017).

Effect of Biochar on Soil Macronutrients

Biochar affects the soil macronutrient characteristics. According to Zhang et al., (2017), twelve biochars of different origin, i.e., walnut shells, corn cobs, corn straws, rice straws, at different pyrolyzed temperatures were applied to soil in pot experiment. The examined parameters were total and available N, P, K and ryegrass growth. The results have shown that biochars decreased the available N, increased the available P, K and increased also the ryegrass biomass (Zhang et al., 2017). Another study indicated the use of biochars produced from different biomasses as absorbents to evaluate their efficiency in nitrate fixation (Yang et al., 2017). Yang et al., explained that the nitrate behavior did not significantly alter after the application of biomass except for biochars derived from bamboo pyrolyzed at high temperatures, which presented relatively high sorption to nitrate.

Decrease of Soil Acidification Using Biochar

Acid soils present low fertility and low crop productivity due to the presence of increased toxicities of aluminum and manganese to crops, deficiencies of phosphorus, calcium potassium and magnesium cations and molybdenium (Dai et al, 2017; Fenn et al., 2006). The addition of biochar improved soil properties and reduced the hazards caused by soil acidification and natural acidic soils. Biochar surface contains many functional groups, e.g., $-COO^-$, $-O^-$, which react with metals such as aluminum and manganese ions and H^+ from the soil solution, alleviating partially soil acidity. Moreover, the nature of biochar is alkaline indicating the increase of soil pH values with its addition to soil (Dai et al, 2017).

Effects on Soil Microbial Activity Using Biochar

Biochar influences microbial biomass and composition due to its unique properties. Biochar porous structure functions as the appropriate environment for microbes' development and colonization. Such microbes are bacteria, actinomycetes and arbuscular mycorrhizal fungi (Agegnehu et al., 2017). Liao et al., (2016) reported the influence of cotton straw biochar on soil microbial community composition and activity in drip-irrigated desert soil planted to cotton. The results have shown an increase in soil microbial biomass, carbon substrate utilization, and enzyme activity related to C and N transformation. A new research (Ming et al., 2016) studied various biochars derived from straw residues and wood chips, and added separately to a paddy soil. Biochars' amendment can alter soil microbial properties and more specifically, increase the amount of general bacteria and consequently increase the concentration of phospholipid fatty acids. Moreover, biochars ameliorated soil nutrient content, affecting the N transformation process. Zhu et al., (2017a) reported that the addition of biochars in soils changed soil properties leading to the modification of microbial activity while certain reactive organic substances and heavy metals contained in biochar may influence negatively soil microorganisms due to toxicity. Muhammad et al., 2014 examined the application of different biochars, derived from swine manure, fruit peels, *Phragmites australis* and *Brassica rapa,* to a sandy loam soil. Different biochars affected microbial communities and total phospholipid fatty acid (PLFA) concentrations. Swine manure, fruit peels, and *Phragmites australis* biochars augmented the bacteria, actinomycetes G-ve bacteria, sulfate reducers, cyclopropane monosaturated and total PLFAs compared to *Brassica rapa* biochars. The enhancement of soil parameters such as pH, C:N ratio, dissolved organic carbon, total carbon and K concentrations affected the microbial communities. A new research study that was applied in two calcareous soils examined the correlation between the corn stalk biochar addition and the activity of several cellular enzymes, which are responsible for C and N cycling and microbial metabolism (Khadem et al., 2017). Biochar may improve the enzyme activities to arid soils and consequently soil quality, carbon sequestration and biogeochemical cycles. A two year maize field experiment took place in silty loam soils on the Loess Plateau in North-Western China where examined the changes in soil microbial biomass carbon, nitrogen, phosphorus, urease and alkaline phosphatase activities with biochar addition (Zhu et al., 2017b). According to the results, biochar increased microbial biomass carbon in most cases, microbial biomass nitrogen and alkaline phosphate activities.

Remediation of Soil and Effluents Using Zeolites

Remediation technologies can be divided in two main categories: in-situ (Breck, 1974) and ex-situ remediation methods (Reddy, et al., 1999) as well as on site and off site technologies. Soil remediation can be performed with easily available additive materials of natural origin found near the place of application (Burlakovs et al., 2012). Control of pollutant receptor linkage can be achieved by utilization materials such as: 1) composts derived from materials such as sewage sludges and other municipal sources, 2) natural or synthetic zeolites, 3) industrial by-products (Gadepalle et al., 2007). These materials create soil conditions that immobilize contaminants while increasing efficient use of fertilizers and water holding capacity. In fact the positive effects of zeolite (clinoptilolite) on biomass and on nutrient status of plant and soils appeared when it was mainly combined with compost or reduced doses of fertilizer (WasterReuse LIFE10 ENV/GR/594, www.wastereuse.eu). The addition of low cost clay minerals such as zeolite and vermiculite with porous structures can lead to the adsorption of macro- and micro-nutrients and in their gradual decrease in soil solution under the appropriate conditions of pH and temperature producing a new biological slow-release fertilizer.

Heavy metal phytoavailability may be reduced if the metals are sorbed or precipitated from the soil solution. The zeolites are widely used for the environmental remediation and restoration. They can decrease bioavailability of heavy metals (Cadmium, Cesium, Chromium, Copper, Lead, Lithium, Mercury and Zinc) in soils and in effluents (Dercova et al., 2007; Indianara et al., 2009; Misaelides, 2011), mainly due to their high CEC (Curkovic et al., 1997). The exchange efficiency depends on the micro-porosity and exchange capacity of the particular zeolite (Nissen et al, 2000). Zeolites can adjust soil pH modestly and do not import new pollution (Mondales et al, 1995; Mohamed et al, 2001; Mozgawa, et al, 2000). They can diminish the solubility of metals by forming salts, oxides, and metal-carbonate precipitates (Chlopecka and Adriano, 1996; Querol, et al., 2006). In addition surface modifications such as heating, surfactant treatment, microwave irradiation etc., were used to enhance sorption capacity or ion-exchange property (Bowman, 2003).

Heavy metals found in leachates posed great threats on soils and natural waters near unlined landfills (Kasassi et al., 2008). High content of organic matter contained in landfill leachate can pollute groundwater and surface water. However, few studies focus on decreasing biotoxicity or eliminating toxic pollutants from landfill leachate. Soil leaching toxicity is an indicator for effectiveness of soil remediation materials (Liu et al., 2015). Zeolites are used for remediations of organic matter polluted soils (Gan, et al., 2009) and organic wastewater (Metes et al., 2004). They appear to be one of the most promising sorbents to remove specific contaminants from aqueous solution (Tashauoei et al., 2010). Natural zeolites are mainly used as ammonium exchangers for wastewater treatment applications (Hedstrom, 2001). Clinoptilolite is known for its ability to remove ammonium from polluted waters (Rahmani and Mahvi, 2006). Liu et al, (2015) showed that natural and microwave modified zeolites increased soil pH, eliminate COD and ammonium in soil extraction which may be attributed to degradation reactions in zeolites relative to sorption reactions as well as decrease leaching toxicity of leachate polluted soils.

Clinoptilolite was most effective in immobilizing lead and cadmium in soils (Gadepalle et al., 2007). Shi et al., (2009) found that the available Pb was reduced by exogenous natural zeolite in the lead-polluted garden soil. Haidouti (1997) reported that the application of the zeolite clinoptilolite at 1–5% (w/w) addition reduced uptake of the Hg in rye grass and alfalfa. Addition of zeolite resulted in a decreased available form of Pb in soil and reduced Pb uptake by rape plants (Li et al, 2009). These were attributed to the fact that zeolite modify soil physical and chemical properties, protect soil organic matter and increase in soil aggregates, increase soil pH and CEC. Nevertheless, a suitable dose of additive zeolite is required for remediation of Pb contaminated garden soil. Ponizovsky and Tsadilas (2003) compared the retention of lead by a sandy loam soil from Greece and the zeolite clinoptilolite. It was concluded that clinoptilolite adsorbed 10–20 times more Pb than soil and sorption by the zeolite was independent of pH. Moirou et al., (2001) conducted column studies that showed that 50–60% of Pb was immobilized with the 20% (w/w) zeolite amendment. Zorpas et al., (2000) used clinoptilolite for the immobilization of heavy metals from sewage sludge, when it was used as a bulking agent during composting. Their results demonstrated that by using 25–30% of clinoptilolite during composting significant amounts of heavy metals Zn, Cd, were adsorbed out of solution. Zeolites, when co-composted with organic wastes may significantly assist in the immobilization of heavy metals due to the fact that they have high CEC and selectivity for heavy metals as well as for several toxic organic substances and thus, increase substantially the quality of the final products.

Synthetic zeolites have also the potential for metal immobilization. The Cd concentration in leaching solution was found to be inversely proportional to the rate of addition of the zeolite synthesized from fly ash (Lin et al., 1998). It was concluded that immobilization of Cd in soil was favoured by the higher cation exchange capacity of the zeolite/soil mixture and higher pH. Gworek (1992) and Edwards et al., (1999) conducted experiments with the synthetic zeolites 4A and 13X in order to determine their effects on the mobility of cadmium. The results have shown a significant reduction in cadmium concentrations in lettuce leaves when compared with the control (Gworek, 1992) as well as reduction in metal content of mobile fraction of Cd (Edwards et al., 1999).

Despite good results for immobilization of metals by zeolites, many works reported that zeolite addition failed to provide satisfactory growth of plants grown in contaminated soils (Nishita et al., 1968; Stead, 2002; Geebelen et al., 2002; Coppola et al., 2003). Excessive zeolite application could induce negative effects on soils such as high alkalinity, which could cause negative influences to soil biological activity. The excess sodium content of pH treated commercial zeolites had a negative effect on crop development despite of advantages in soil moisture holding and cation exchange properties (Ramesh et al, 2015). Geebelen et al., (2002) observed significant reductions in the root growth and leaf area independent of soil Pb concentration. This was ascribed to the fact that zeolite had negatively influence soil physical properties where a dispersed soil structure was evident due to the addition of sodium ions. Coppola et al., (2003) reported a negative effect on wheat growth due to high amendment rates of zeolite to soils. Addition of 50% and 75% zeolite (w/w), there was a strong decrease in the dry plant biomass and almost no plants survived in the pots with a 75% conditioner addition. This was attributed to the increase in the electrical conductivity and Na toxicity. Nishita et al., (1968) observed a 10% decrease in the yield of bean when Na-clinoptilolite was used as a soil amendment. This was due to high Na concentration in the soil, causing osmotic and toxicity problems. Stead (2002) also reported a similar negative effect on growth of rye grass and clover which were grown in pots with 5%, 10% 20% addition of a sodium-zeolite. This impact was due to high sodium exchanged from the zeolite. Therefore application dosages would act as an essential factor in remediation process (Liu et al, 2015).

It can be concluded that natural zeolites can immobilize metals due to their high ion exchange capacities and highly porous structures. In addition synthetic zeolites have been reported to be effective at reducing soil phytotoxicity and improving plant growth. Zeolite used as a soil conditioner can reduce the need for fertilizer whilst improving plant growth. (Gadepalle et al., 2007). On the other hand, care should be taken when using zeolites with high sodium contend. Despite promising results for immobilization of contaminants by zeolites, several studies have demonstrated that zeolites addition failed to promote satisfactory growth of plants grown in contaminated soils. Thus, of serious concern within contaminated land remediation should be the possibility of 'Na toxicity' to the surrounding soil, and hence uptake by plants. It has been shown that they increase soil salinity and exchangeable sodium concentrations, which could cause plant toxicity.

CONCLUSION

Based on the above scientific reports, the basic conclusions of biochar or zeolite application are the following:

- Biochar characteristics vary according to the raw biomass materials, heating temperature, heating time and conditions of biochars' production
- Biochars are highly suitable as soil amendments of contaminated soils with organic and inorganic compounds
- They influence microbial biomass and composition due to their unique porous structure
- They provide essential nutrients to soils improving crop yields and food quality

- Zeolites immobilize metals due to their porous structure and ion exchange capacity
- Biochars and zeolites modify soil physical and chemical properties
- Excessive zeolite application failed to provide satisfactory growth of plants in contaminated soils

REFERENCES

Agegnehu, G., Srivastava, A. K., Bird, M. I., 2017. The role of biochar and biochar – compost in improving soil quality and crop performance: A review, *Applied Soil Ecology*, 119, 156-170.

Bowman, R. S., 2003. Applications of surfactant-modified zeolites to environmental remediation, *Microporous and Mesoporous Materials*, 61, 43-56.

Breck, D. W., 1974. *Zeolite Molecular Sieves: Structure, Chemistry and Use.* New York: John Wiley & Sons.

Burlakovs, J., Klavins, M., Karklina, A., 2012. Remediation of soil contamination with heavy metals by using zeolite and humic acid additives. *Latvian Journal of Chemistry*, 4, 336–341.

Cao, Y., Yang, B., Song, Z., Wang, H., He, F., Han, X., 2016. Wheat straw biochar amendments on the removal of polycyclic aromatic hydrocarbons (PAHs) in contaminated soil, *Ecotoxicology and Environmental Safety*, 130, 248-255.

Cárdenas-Aguiar, E., Gascó, G., Paz-Ferreiro, J., Méndez, A., 2017. The effect of biochar and compost from urban organic waste on plant biomass and properties of an artificial copper polluted soil, *International Biodeterioration & Biodegradation* 124, 223-232.

Chlopecka, A., Adriano, D. C., 1996. Mimicked in-situ stabilization of metals in a cropped soil: bioavailability and chemical form of zinc. *Environmetal Science & Technology,* 30, 3294–3303.

Coppola, E. I., Battaglia, G., Bucci, M., Ceglie, D., Colella, A., Langella, A., Buondonno, A., Colella, C., 2003. Remediation of Cd- and Pb-polluted soil by treatment with organozeolite conditioner. *Clays and Clay Minerals*, 51, 609–615.

Curkovic, S., Cerjan-Stefanovic, S., Filipan, T., 1997. Metal ion exchange by natural and modified zeolites. *Water Research,* 31, 1379–1382.

Dai, Z., Zhang, X., Tang, C., Muhammad, N., Wu, J., Brookes, P. C., Xu, J., 2017. Potential role of biochars in decreasing soil acidification- A critical review, *Science of the Total Environment*, 581-582, 601-611.

Dercova, K., Sejakova, Z., Skokanova, M., Barancikova, G., Makovnikova, J., 2007. Bioremediation of soil contaminated with pentachlorophenol using humic acids bound on zeolite. *Chemosphere,* 66, 783–790.

Edwards, R., Rebedea, I., Lepp, N. W., Lovell, A. J., 1999. An investigation into the mechanism by which synthetic zeolites reduce labile metal concentrations in soils, *Environmental Geochemistry & Health*, 21, 157–173.

Fenn, M. E., Huntington, T., McLaughlin, S., Eagar, C., Gomez, A., Cook, R., 2006. Status of soil acidification in North America, *Journal of Forest Science* 52, 3-13.

Gadepalle, V. P., Ouki S. K., Van Herwijnen R., Hutchings T., 2007. Immobilization of heavy metals in soil using natural and waste materials for vegetation establishment on contaminated sites, *Soil and Sediment Contamination: An International Journal*, 16(2), 233-251.

Gan, S., Lau, E. V., Ng H. K., 2009. Remediation of soils contaminated with polycyclic aromatic hydrocarbons (PAHs). *Journal of Hazardous Materials*, 172, 532-549.

Geebelen, W., Vangronsveld, J., Adriano, D. C., Carleer, R., Clijsters, H., 2002. Amendment induced immobilization of lead in a lead spiked soil: Evidence from phytotoxic studies. *Water Air & Soil Pollution,* 140, 261–277.

Glaser, B., Haumaier, L., Guggenberger, G., Zech, W. 2001. The 'Terra Preta' phenomenon: a model for sustainable agriculture in the humid tropics, *Naturwissenschaften*, 88 (1) 37–41.

Gu, J., Zhou, W., Jiang, B., Wang, L., Ma, Y., Guo, H., Schulin, R., Ji, R., Evangelou, M. W. H., 2016. Effects of biochar on the transformation and earthworm bioaccumulation of organic pollutants in soil, *Chemosphere*, 145, 431-437.

Gworek, B., 1992. Inactivation of cadmium in contaminated soils using synthetic zeolites. *Environmental Pollution,* 75, 269–271.

Haidouti, C., 1997. Inactivation of mercury in contaminated soils using natural zeolites. *Science of the Total Environment,* 208, 105–109.

Hedström, A., 2001. Ion exchange of ammonium zeolites: A literature review. *Journal of Environmental Engineering*, 127(8), 673-681.

Inglezakis, V.J., Zorpas, A. A., 2012. *Handbook of Natural Zeolites*, DOI: 10.2174/97816080526151120101

International Biochar Initiative 2017. http://www.biochar-international.org/biochar.

International Biochar Initiative 2013. *Standardized product definition and product testing guidelines for biochar that is used in Soil*, pp. 1-48, http://www.biochar-international.org/sites/default/files/IBI_Biochar_Standards_V1.1.pdf

Kambo, H. S., Dutta, A., 2015. A comparative review of biochar and hydrochar in terms of production, physico-chemical properties and applications, *Renewable and Sustainable Energy Reviews*, 45, 359-378.

Kasassi, A., Rakimbei, P., Karagiannidis, A., Zabaniotou, A., Tsiouvaras, K., Nastis, A., Tzafeiropoulou, K., 2008. Soil contamination by heavy metals: Measurements from a closed unlined landfill, *Bioresource Technology*, 99, 8578-8584.

Kulasekaran, R., Damodar Reddy, D., 2011. Zeolites and Their Potential Uses in Agriculture, *Advances in Agronomy*, 113, 215-236.

Khadem A., Raiesi F., 2017. Influence of biochar on potential enzyme activities in two calcareous soils of contrasting texture, *Geoderma*, 308, 149-158.

Khalid, S., Shahid, M., Niazi, N. K., Murtaza, B., Bibi, I., Dumat, C., 2017. A comparison of technologies for remediation of heavy metal contaminated soils, *Journal of Geochemical Exploration*, 182, 247-268.

Koltowski M., Hilber I., Bucheli T. D., Charmas B., Skubiszewska-Zięba J., Oleszczuk P., 2017. Activated biochars reduce the exposure of polycyclic aromatic hydrocarbons in industrially contaminated soils, *Chemical Engineering Journal*, 310, 33-40.

Koltowski, M., Hilber, I., Bucheli, T. D., Oleszczuk, P., 2016. Effect of steam activated biochar application to industrially contaminated soils on bioavailability of polycyclic aromatic hydrocarbons and ecotoxicity of soils, *Science of the Total Environment*, 566-567, 1023-1031.

Liao, N., Li, Q., Zhang, W., Zhou, G., Ma, L., Min, W., Ye J., Hou, Z., 2016. Effects of biochar on soil microbial community composition and activity in drip-irrigated desert soil, *European Journal of Soil Biology* 72, 27-34.

Liu, J., Ding, Y., Ma, L., Gao, G., Wang, Y., 2017. Combination of biochar and immobilized bacteria in cypermethrin-contaminated soil remediation, *International Biodeterioration & Biodegradation*, 120, 15-20.

Lone, A. H., Najar, G. R., Ganie, M. A., Sofi, J. A., Ali, T., 2015. Biochar for sustainable soil health: A review of prospects and concerns, *Pedosphere*, 25(5), 639-653.

Lu, K., Yang, X., Gielen, G., Bolan, N., Ok, Y.S., Niazi, N.K., Xu, S., Yuan, G., Chen, X., Zhang, X., Liu, D., Song, Z., Liu, X., Wang, H., 2017. Effect of bamboo and rice straw biochars on the mobility and redistribution of heavy metals (Cd, Cu, Pb and Zn) in contaminated soil, *Journal of Environmental Management,* 186, 285-292.

Li, H., Shi, W., Shao, H., Shao, M., *2009*. The remediation of the lead-polluted garden soil *by* natural zeolite*, Journal of Hazardous Materials,* 169, 1106–1111.

Lin, C. F., Lo, S., Lin, H., and Lee, Y. 1998. Stabilization of cadmium contaminated soils using synthetic zeolites, *Journal of Hazardous Materials*, 60, 217–226.

Liu, Y., Chen, Z., Yang, L., Liao, Z., Sun, T., Wang, J., Wang, H., 2015. Short-term remediation of soil contaminated with landfill leachate using natural and microwave modified zeolite of the clinoptilolite type. *Bulgarian Chemical Communications*, 47(D), 193 – 200.

Metes, A., Kovacevic, D., Vujevic, D., Papic, S., 2004. The role of zeolites in wastewater treatment of printing inks. *Water Research*, 38, 3373-3381.

Ming, D. W., Boettinger, J. L. Zeolites in soil environments. In: Bish D. L. and Ming D. W. (ed.) Natural Zeolites: Occurrence, Properties, Applications. Reviews in Mineralogy and Geochemistry, Washington, DC; *Mineralogical Society of America and Geochemical Society*; 2001; 323-345.

Ming, L., Ming, L., Zhong-pei, L., Chun-yu, J., Meng, W., 2016. Soil N transformation and microbial community structure as affected by adding biochar to a paddy soil of subtropical China, *Journal of Integrative Agriculture* 15(1), 209–219.

Misaelides, P., 2011. Application of natural zeolites in environmental remediation: A short review, *Microporous and Mesoporous Materials, 144(1-3), 15-18.*

Moirou, A., Xenidis, A., Paspaliaris, I. 2001. Stabilization Pb, Zn, and Cd-contaminated soil by means of natural zeolite. *Soil and Sediment Contamination: An International Journal*. 10(3), 251–267.

Mondale, K. D., Carland, R. M., Aplan, F. F., 1995. The comparative ion exchange capacities of natural sedimentary and synthetic zeolites, *Minerals Engineering,* 8(4-5), 535–548.

Mohamed, M. M., 2001. Heat capacities, phase transitions and structural properties of cation-exchanged H-mordenite zeolites, *Thermochimica Acta,* 372, 75–83.

Mozgawa, W., 2000. The influence of some heavy metals cations on the FTIR spectra of zeolites, *Journal of Molecular Structure,* 555(1-3), 299–304.

Moore F., Gonzalez M. E., Khan N., Curaqueo G., Sanchez-Monedero M., Rilling J., Morales E., Panichini M., Mutis A., Jorquera M., Mejias J., Hirzel J., Meier S., 2017. Copper immobilization by biochar and microbial community abundance in metal-contaminated soils, *Science of the Total Environment*, (in press).

Morillo, E., Villaverde, J., 2017. Adsvanced technologies for the remediation of pesticide-contaminated soils, *Science of the Total Environment*, 586, 576-597.

Muhammad N., Dai Z., Xiao K., Meng J., Brookes P. C., Liu X., Wang H., Wu J., Xu J., 2014. Changes in microbial community structure due to biochars generated from different feedstocks and their relationships with soil chemical properties, *Geoderma*, 226-227, 270-278.

Nishita, H., Haug, R. M., 1971. Influence of clinoptilolite on Sr^{90} and Cs^{137} uptakes by plants. *Soil Science,* 114(2), 149–157.

Nissen, L. R., Lepp, N. W., Edwards, R., 2000. Synthetic zeolites as amendments for sewage sludge-based compost, *Chemosphere*, 41, 265–269.

Oliveira, F. R., Patel, A. K., Jaisi, D. P., Adhikari, S., Lu, H., Khanal, S. K., 2017. Environmental application of biochar: Current status and perspectives, *Bioresource Technology*, 246, 110-122.

Ostroski, I. C., Barros, M. A., Silva, E. A., Dantas, J. H., Arroyo, P. A., Lima, O. C., 2009. A comparative study for the ion exchange of Fe(III) and Zn(II) on zeolite NaY, *Journal of Hazardous Materials*, 161, 1404–1412.

Ponizovsky, A. A., Tsadilas, C. D., 2003. Lead(II) retention by Alfisol and clinoptilolite: cation balance and pH effect, *Geoderma*, 5, 303–312.

Pukalchik, M., Merci, F., Panova, M., Břendová, K., Terekhova, V. A., Tlustoš, P., 2017. The improvement of multi-contaminated sandy loam soil chemical and biological properties by the biochar, wood ash and humic substances amendments, *Environmental Pollution*, 229, 516-524.

Qiao, Y., Wu, J., Xu, Y., Fang, Z., Zheng, L., Cheng, W., Tsang, E. P., Fang, J., Zhao, D., 2017. Remediation of cadmium in soil by biochar – supported iron phosphate nanoparticles, *Ecological Engineering* 106, 515-522.

Qiu, Y., Zheng, Z., Zhou, Z., Sheng, G. D., 2009. Effectiveness and mechanisms of dye adsorption on a straw-based biochar, *Bioresource Technology*, 100, 5348-5351.

Querol, X., Alastuey, A., Moreno, N., Alvarez-Ayuso, E., Garcia-Sanchez, A., Cama, J., Ayora, C., Simon, M., 2006. Immobilization of heavy metals in polluted soils by the addition of zeolite material synthesized from coal fly ash, *Chemosphere*, 62, 171–180.

Rahmani, A. R., Mahvi, A. R., 2006. Use of ion exchange for removal of ammonium: A biological regeneration of zeolite, *Global NEST Journal*, 8, 146-150.

Ramesh, V., George, J., Jyothi, J.S., Shibli, S.M.A., 2015. Effect of zeolites on soil quality, plant growth and nutrient uptake efficiency in sweet potato (*Ipomoea batatas L.*), *Journal of Root Crops*, 41(1), 25-31

Rao, M. A., Simeone, G. D. R., Scelza, R., Conte, P., 2017. Biochar based remediation of water and soil contaminated by phenanthrene and pentachlorophenol, *Chemosphere*, 186, 193-201.

Reddy, K. R., Adams, J. A., Richardson, C., 1999. Potential technologies for remediation of Brownfield. *Practice Periodical of Hazardous, Toxic, and Radioactive Waste Management*, 3(2), 61–68.

Shen, Z., Hou, D., Zhao, B., Xu, W., Ok, Y. S., Bolan, N. S., Alessi, D. S., 2018. Stability of heavy metals in soil washing residue with and without biochar addition under accelerated ageing, *Science of the Total Environment*, 619-620, 185-193.

Shi, W. Y., Shao H. B., Li H., Shao M. A., Du S., 2009. Progress in the remediation of hazardous heavy metal-polluted soils by natural zeolite. *Journal of Hazardous Materials*, 170(1), 1-6.

Simitzis, J., Ioannou, Z. 2012. Activated carbonaceous materials based on thermosetting binder precursors, Chap. 12, *Activated carbon*, ed. J. F. Kwiatkowski, Nova Science Publishers, Inc.

Stead, K., 2002. *Environmental implications of using the natural zeolite clinoptilolite for the remediation of sludge amended soils*. PhD thesis, University of Surrey, United Kington.

Su, H., Fang, Z., Tsang, P. E., Zheng, L., Cheng, W., Fang, J., Zhao, D., 2016. Remediation of hexavalent chromium contaminated soil by biochar-supported zero-valent iron nanoparticles, *Journal of Hazardous Materials* 318, 533-540.

Tang, J., Zhu, W., Kookana, R., Katayama, A., 2013. Characteristics of biochar and its application in remediation of contaminated soil, *Journal of Bioscience and Bioengineering* 116(6), 653-659.

Tan, X., Liu, Y., Zeng, G., Wang, X., Hu, X., Gu, Y., Yang, Z., 2015. Application of biochar for the removal of pollutants from aqueous solutions, *Chemosphere*, 125, 70-85.

Tashauoei, H. R., Attar Movahedian, H., Amin, M. M., Kamali, M., Nikaeen, M., Vahid Dastjerdi, M., 2010. Removal of cadmium and humic acid from aqueous solutions using surface modified nanozeolite A. *International Journal of Environmental Science and Technology*, 7(3), 497-508.

Van Poucke, R., Ainsworth, J., Maeseele, M., Ok, Y. S., Meers, E., Tack, F. M. G., 2017. Chemical stabilization of Cd-contaminated soil using biochar, *Applied Geochemistry* (in press).

Vithanage M., Herath I., Joseph S., Bundschuh J., Bolan N., Ok Y. S., Kirkham M. B., Rinklebe J., 2017. Interaction of arsenic with biochar in soil and water: A critical review, *Carbon* 113, 219-230.

Wahi, R., Zuhaidi, N. F. Q., Yusof, Y., Jamel, J., Kanakaraju, D., Ngaini, Z. 2017. Chemically treated microwave-derived biochar: An overview, *Biomass and Bioenergy* 107, 411-421.

Wang Y., Xu Y., Li D., Tang B., Man S., Jia Y., Xu H., 2017. Vermicompost and biochar as bio-conditioners to immobilize heavy metal and improve soil fertility on cadmium contaminated soil under acid rain stress, *Science of the Total Environment*, (in press).

Yang, J., Li, H., Zhang, D., Wu, M., Pan, B., 2017. Limited role of biochars in nitrogen fixation through nitrate adsorption, *Science of the Total Environment*, 592, 758-765.

Yang, Z., Fang, Z., Tsang, P. E., Fang, J., Zhao, D., 2016b. In situ remediation and phytotoxicity assessment of lead contaminated soil by biochar – supported nHAP, *Journal of Environmental Management*, 182, 247-251.

Yang, Z., Fang, Z., Zheng, L., Cheng, W., Tsang, P. E., Fang, J., Zhao, D., 2016a. Remediation of lead contaminated soil by biochar – supported nano-hydroxyapatite, *Ecotoxicology & Environmental Safety*, 132, 224-230.

Yoo, J.-C., Beiyuan, J., Wang, L., Tsang, D. C. W., Baek, K., Bolan, N. S., Ok, Y. S., Li, X.-D., 2017. A combination of ferric nitrate/EDDS-enhanced washing and sludge-derived biochar stabilization of metal-contaminated soils, *Science of the Total Environment* (in press).

Zhang, G., Guo, X., Zhao, Z., He, Q., Wang, S., Zhu, Y., Yan, Y., Liu, X., Sun, K., Zhao, Y., Qian, T., 2016. Effects of biochars on the availability of heavy metals to ryegrass in an alkaline contaminated soil, *Environmental Pollution*, 218, 513-522.

Zhang, G., Guo, X., Zhu, Y., Han, Z., He, Q., Zhang, F., 2017. Effect of biochar on the presence of nutrients and ryegrass growth in the soil from an abandoned indigenous coking site: The potential role of biochar in the revegetation of contaminated site, *Science of the Total Environment*, 601-602, 469-477.

Zhu X., Chen B., Zhu L., Xing B., 2017a. Effects and mechanisms of biochar-microbe interactions in soil improvement and pollution remediation: A review, *Environmental Pollution* 227, 98-115.

Zhu L.-X., Xiao Q., Cheng H.-Y., Shi B.-J., Shen Y.-F., Li S.-Q., 2017b. Seasonal dynamics of soil microbial activity after biochar addition in a dryland maize field in North-Western China, *Ecological Engineering*, 104, 141-149.

Zorpas, A. A., Kapetanios, E., Zorpas, G. A., Karlis, P., Vlyssides, A., Haralambous, I., Loizidou, M., 2000. Compost produced from organic fraction of municipal solid waste, primary stabilized sewage sludge and natural zeolite, *Journal of Hazardous Materials*, 77, 149–159.

INDEX

D

E

F

G

H

I

K

L

M

N

O

P

Q

R

S

T

U

V

W

X

Z